T0214708

Synthesis Lectures on Engineering, Science, and Technology

The focus of this series is general topics, and applications about, and for, engineers and scientists on a wide array of applications, methods and advances. Most titles cover subjects such as professional development, education, and study skills, as well as basic introductory undergraduate material and other topics appropriate for a broader and less technical audience.

Sanjay Kumar

Additive Manufacturing Classification

 Springer

Sanjay Kumar
Gumla, India

ISSN 2690-0300   ISSN 2690-0327   (electronic)
Synthesis Lectures on Engineering, Science, and Technology
ISBN 978-3-031-14222-2   ISBN 978-3-031-14220-8   (eBook)
https://doi.org/10.1007/978-3-031-14220-8

This Springer imprint is published by the registered company Springer Nature Switzerland AG
The registered company address is: Gewerbestrasse 11, 6330 Cham, Switzerland

Contents

About the Author

Dr. Sanjay Kumar is the author of *Additive Manufacturing Processes*, 2020, Springer, and *Additive Manufacturing Solutions*, 2022, Springer, Cham (https://www.springer.com/gp/book/9783030807825).

Abbreviations

2PP	Two-photon Polymerization
3DGP	3D Gel-Printing
AFSD	Additive Friction Stir Deposition
AJ	Aerosol Jetting
ALM	Additive Layer Manufacturing
AM	Additive Manufacturing
ANLM	Additive Non-Layer Manufacturing
BJ3DP	Binder Jet Three-Dimensional Printing
CAD	Computer-Aided Design
CLF	Ceramic Laser Fusion
CLIP	Continuous Liquid Interface Production
CNC	Computer Numerical Control
CSAM	Cold Spray AM
DW	Direct Writing
E-beam	Electron Beam
EBM	Electron Beam Melting
ECAM	Electrochemical AM
EPBF	Electron beam Powder Bed Fusion
ESD	Electron beam Solid Deposition
FDC	Fused Deposition of Ceramics
FDM	Fused Deposition Modeling
FGM	Functionally Graded Material
FPM	Fused Pellet Modeling
FSBAM	Friction Surfacing based AM
FSP	Friction Stir Processing
GMAW	Gas Metal Arc Welding
GTAW	Gas Tungsten Arc Welding
HSS	High Speed Sintering
IJP	Ink Jet Printing
IM	Injection Molding

L-beam	Laser beam
LD	Liquid Deposition
LENS	Laser Engineered Net Shaping
LMD	Laser Metal Deposition
LMHAM	Localized Microwave Heating based AM
LPBF	Laser Powder Bed Fusion
LSD	Laser Solid Deposition
MAPS	Microheater Array Powder Sintering
MDDM	Micro Droplet Deposition Manufacturing
MJF	Multi Jet Fusion
PB	Powder Bed
PBF	Powder Bed Fusion
PBP	Powder Bed Process
PJ	Photopolymer Jetting
PPBP	Photopolymer Bed Process
RFP	Rapid Freeze Prototyping
SD	Solid Deposition
SHS	Selective Heat Sintering
SIS	Selective Inhibition Sintereing
SL	Stereolithography
SLM	Selective Laser Melting
SLS	Selective Laser Sintering
STL	Standard Triangle Language
T3DP	Thermoplastic 3D Printing
WAAM	Wire Arc AM
WEAM	Wire Electron beam AM

Process and Classification

1

1   Introduction

Additive process implies that materials are added to make a part. Examples are sintering, casting, injection molding (IM), stereolithography (SL), ink jet printing (IJP), selective laser sintering (SLS), selective laser melting (SLM), laser metal deposition (LMD), etc. These processes differ from subtractive processes as the latter do not add but remove the material. Examples of subtractive processes are drilling, boring, milling, sawing, electrical discharge machining, laser ablation, water jet cutting, grinding, etc.

Additive processes differ from deforming processes as the latter deform the material. Examples of deforming processes are deep drawing, stamping, incremental forming, bending, forging, etc. [1].

Additive processes are divided into two groups. The first group consists of sintering, casting, IM while the second consists of SL [2], IJP [3], SLS [4], electron beam melting (EBM) [5], LMD [6], etc. These examples show that the first group requires design-specific tooling to make a part while the second does not. The processes belonging to the second group are additive manufacturing (AM), which exist for last few decades, but the concept of AM exists since antiquity [7]. AM is also known as 3D printing [8].

2   Role of Design-Specific Tool

In IM (a non-AM additive process), a mold is required to make a part of one design. A number of parts can be made, but all must have the same design. If a part of another design is required, a mold of another design is required. Thus for making parts of different designs, different molds are required.

© The Author(s), under exclusive license to Springer Nature Switzerland AG 2022   1
S. Kumar, *Additive Manufacturing Classification*, Synthesis Lectures on Engineering, Science, and Technology, https://doi.org/10.1007/978-3-031-14220-8_1

In LMD (an AM process), a part is made by depositing material along a certain track as per design. If a part of another design is required, the material is deposited along another track as per new design—it is the pattern of deposition that determines the types of designs to be made. Thus material is deposited differently, and a new part having new design is formed. Therefore, a number of different parts can be made without having the need for different design-specific molds.

This is how LMD differs from IM, or how the role of a design-specific tool is different in AM and non-AM additive process.

3   Tools in AM and Other Processes

Some AM processes require tools. For example, additive friction stir deposition (AFSD) [9], an AM process, requires tools to accomplish the process. Then, how the requirement of tools in AM is different from that in a non-AM additive process, e.g., IM?

In AFSD, if the dimension of a design is big, one tool will wear out before it makes a complete part, more tools will be required to complete the task. If some features of the design are small, the tool of the same size will be oversized and be unable to make the part without compromising its accuracy. Hence, a small tool is required for making a small feature.

In IM, the design of a part is the same as the design of the cavity of a mold while in AFSD, the design of a part does not need to match the shape of a tool. Thus in AFSD, tools are required so fabrication as per any design will be completed with the help of either one or more tools while in IM, molds are required so the fabrication as per a particular design will be completed with the help of a particular mold.

In AFSD, a tool is a rod or pin used to convert powders to non-powder solid material using friction energy. The conversion is the primary aim of the tool. The second aim is to create a well-defined track along with the conversion. The conversion is primary because if there is no conversion, no part will be formed, and if the conversion is partial, a part having incomplete property will be formed. But, the conversion itself is not adequate, it must lead to a well-defined track that on adding up will give a well-defined part. Otherwise, the part will require post-processing to overcome the demerit of an inadequate conversion, as is done in AM [10].

For making a small feature, e.g. a thin wall, a small tool that does not make a wider track than the width of the wall is required while for a big feature, a big tool is required as the same small tool will either decrease the fabrication speed or break due to overstrain.

Thus, if a design does not consist of a small or big feature, one tool is sufficient for converting the design into a part; otherwise, two tools are required. If the design becomes more complex and consists of many features of sizes far different from each other, more tools are required. This situation reminds of milling, where tools of various sizes are needed to remove materials to convert a complex design into a part.

Thus, AFSD does not have an edge over milling for fulfilling the goal of toolless manufacturing. Though the purpose of tools in milling is in sharp contrast to that in AFSD. If milling uses tools for subtraction, AFSD uses them for addition.

Considering another AM where tools are used. SLM uses two lasers for processing: one of big diameter for making big features or speed up production, and other is of small diameter for making small features. In this example, SLM can be considered to be using two tools for processing, one tool is an l-beam of big diameter while the other tool is of small diameter.

Then, why SLM should be considered toolless manufacturing while machining should not? This is because the tool of SLM unlike a machining tool does not wear out while interacting with material. Consequently, the performance of SLM is not restricted because of the loss of its tool caused by tool–material interaction while machining performance is restricted by the wear of its tool due to the interaction. Thus, tool–material interactions have ability to degrade the performance of machining but not of SLM, making SLM relatively toolless manufacturing when comparing it with machining. Though the tool of SLM, which is a laser beam, degrades. But, it degrades not because it interacts with a hard material but because it has a specific life-time irrespective of the types of materials it interacts with. In machining, the degradation of a tool depends upon the type of materials it interacts with, making its tool unlike SLM tool not independent of the materials.

There can be a question: why SLM is relatively toolless manufacturing when comparing it with machining and why not instead it is absolutely toolless manufacturing. It is because: though, SLM is free from traditional tools used in manufacturing, it is not free from the concept of tools [11].

What if tool in a particular AM starts to wear in the same way a tool wears in traditional manufacturing? Then, this particular AM is not toolless manufacturing, e.g. AFSD. But AFSD, being not toolless manufacturing, does not mean it does not have advantage over machining that is also not toolless manufacturing. It implies this advantage does not make AFSD different from machining when compared on the basis of not being toolless manufacturing.

4   Definition of a Process and AM

Oxford dictionary gives the definition of a process as 'a series of actions or steps taken in order to achieve a particular end'. AM processes are a combination of various steps in sequence. Such steps, for example, are: what are materials, how they are brought, how they are converted. For example, three steps in powder bed fusion (PBF), which defines the process, are: materials are in the form of a powder, they are placed in the form of a bed, and they are converted by fusion.

Processes that come under PBF follow the same steps but differ in the sources responsible for fusing them. For example, EBM and SLM are two PBF, which differ because they are fused by e-beam and l-beam, respectively. SLS and SLM are two PBF, which are fused by the same type of applied energy source, i.e. l-beam but differ because both use the l-beam for different purposes. SLS uses it for partial melting while SLM for full melting. Though, the name SLS does not convey the partial melting [12].

All these three PBF, i.e. SLS, SLM and EBM differ in the last step of the process. There is no example of PBF that differs in the first or second step. It is not impossible that a PBF emerges, which differs in the first step such as polymer, metal, ceramic, or composite based, or differs in the second step: porous, curved, or two beds, etc. For example, there emerges a future process 'PBF for polymer', the process steps in this process are so much detailed that the process does not work with any material other than polymers.

A process is executed in a machine or system (machine and system are used inter-changeably in AM). Some machines execute more than one process while some process can be executed in more than one machine. For example, an SLM machine can execute both processes SLM and SLS. By decreasing laser power and preventing full melting, SLS can be achieved. Similarly, an SLS machine can execute both processes SLS and SLM. By choosing low melting point materials such as tin or bronze and facilitating full melting, SLM can be achieved. Thus a process (SLS or SLM) can be executed in two machines (SLS and SLM).

Most of AM machines are process-specific and therefore execute single AM process each. However, if other non-AM process can be included to understand the difference between a process and machine, these AM machines can execute more than one process. For example, LMD machine can make not only a part but also does laser cladding on another part, and thus executes two processes—LMD and laser cladding.

Most of the AM processes have similar beginning and respective AM machines can execute two processes—one their initial process and the other their existing process. For example, wire arc AM (WAAM) is developed from arc welding; WAAM machine can execute two processes—arc welding (initial process) and WAAM (existing process). Other examples are: IJP is developed from paper printing, wire e-beam AM (WEAM) from electron welding, plasma arc AM from plasma welding, friction stir AM from friction stir welding, cold spray AM from cold spray, and electrochemical AM from electro-printing.

5   Steps in Additive Layer Manufacturing (ALM)

In ALM, a part is visualized as an assembly of layers and, therefore, effort is done to make or have layers and then assemble them. When a part needs to be fabricated, its CAD model needs to be transformed in the form of layers (Fig. 1a). Before it is transformed (sliced), the model needs to be converted into an assembly of many small interconnected triangles, which is called tessellation that generates STL (standard triangle language or stereolithography) files.

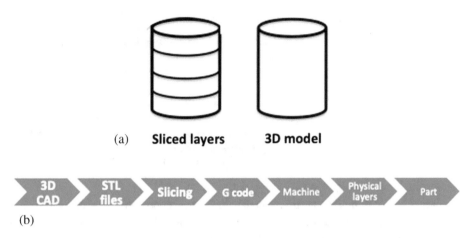

Fig. 1 Conversion of a CAD model into a part: **a** slicing of a cylindrical model into layers, **b** sequence of transformation

Triangles will not be able to exactly match the boundaries of the model if the boundary is not a straight line. Therefore, in the case of a curvature, there remains a gap between the boundary between the CAD and STL model. To slice the STL model, an horizontal plane intersects it at various points and collects data at these points. The separation between two consecutive points is equal to layer thickness [13].

These data determine tool path—information is collected by machine in the form of G-code, which is required by the machine to use its tool accordingly and convert these data (virtual layers) into physical layers. The sequence of steps is given in Fig. 1b. The lowest layer is fabricated first followed by the fabrication of successive layers over it till the topmost layer is fabricated.

6　　Layerwise and Non-layerwise

There are some AM which are not fabricated layerwise. They are non-layerwise or Additive non-layer manufacturing (ANLM). Thus, AM is of two types: (1) ALM, and (2) ANLM (Fig. 2). Two-photon polymerization (2PP) [14] is an example of ANLM.

7　　Machining and AM

7.1　　Approach

Machining is a big-small approach where a big block becomes a small product while AM is a small-big approach where small blocks become a big product. These are two different

Fig. 2 Classification on the
basis of layer formation

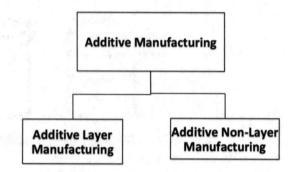

approaches. This brings a question whether these two approaches are still different if observed through the movement of tools. Machining can be a top–down approach while AM can be a bottom–up approach—these approaches can be due to the direction of tool movement—tool moves from top to down in machining while from bottom to up in AM (Fig. 3).

In machining, an object is made by machining a block from the top surface. The object is continuously carved out from the block so that it is continuously getting visible from the top surface to its down surface with the progress of machining. This is shown from Fig. 3a to c through 3b where a block is cut by a machining tool.

This definition does not imply that the machining has limitations and cannot be performed from a side or bottom surface. If machining is performed from the side or bottom surface, these surfaces will act as a top surface for its tool. Top–down approach implies that the tool will go towards the material (block) so that it can access it to machine while

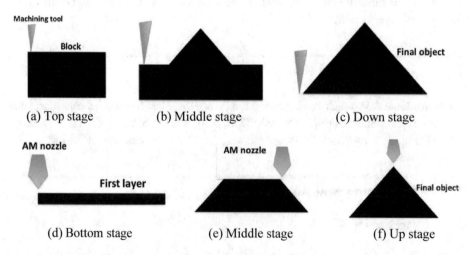

Fig. 3 Top–down approach in machining: **a** starting point, **b** middle stage, **c** final stage. Bottom–up approach in AM: **d** starting point, **e** middle stage, **f** final stage

bottom–up approach implies that the tool will go away from the material (substrate) so that it can have a free space to build.

In AM, a build progresses from a bottom surface or the first layer and reaches the last layer or the top surface. The bottom–up approach tells that it starts from the first layer, adds continuously layers and finishes at the last layer. If the orientation of a CAD model changes by 180° so the top surface becomes bottom surface and vice versa, the approach does not change—it again needs to start from the first layer that then becomes the bottom surface of the build. If the orientation of the AM system changes by 180° (e.g. inverse stereolithography) so the approach may look similar to top–down approach of the machining, the approach remains unchanged—it again needs to start from the first layer that then becomes the bottom surface of the build.

In AM, irrespective of the orientation of the final object or the system, the build starts from nothing, adds building blocks until it gets everything in the form of a final object—in bottom–up approach, fabrication happens through addition. In machining, the build starts from everything, removes blocks until there is nothing left to be removed except a shaped block in the form of a final object—in top–down approach, fabrication happens through disintegration.

Figure 3d–f show the first stage, middle stage, and last stage of the build. These figures show an AM nozzle that is not common to all AM processes but the approach is common. For processes, which do not use an AM nozzle, direction of a build increment replaces the direction of the movement of a tool.

The name 'bottom–up' does not intend to imply the limitation of AM. Considering a case when fabrication happens on the side surface of a vertical wall [15], then the nozzle will not move from bottom to up but either from left to right or from right to left— the name of the approach will be 'left–right' or 'right–left' approach. These new names will not represent anything differently than what is represented by 'bottom–up' approach. If AM is 'left–right' instead of 'bottom–up', machining will be 'right–left' instead of 'top–bottom', and vice versa as the purpose of the name is to show the relative position between AM and machining.

If a vertical wall is standing in the left corner on a horizontal table, the fabrication on its side surface will be towards the right direction. Thus, AM tool moves from left to right in a 'left–right' approach. If instead the machining is performed on the same surface, the tool will move further left so that the tool will pierce inside the surface, the movement of the machining tool will from left to further left in a 'right–left' approach.

This brings a question—if AM tool movement can be explained by 'bottom–up', or 'left–right' or 'right–left' approach, then which name should be preferred. The answer of this question lies in another question—which is more important: symbol or symbolism? If symbol is more important, 'bottom–up' is preferred as it represents working of most of the AM systems. If symbolism is more important, any name will suffice.

The top–down or bottom–up approach is about the tool movement and does not indicate the orientation of AM systems or final AM products formed [16, 17]. Approaches for getting different orientations of an AM product on a substrate are given in Chap. 2.

7.2   Generation of Waste

While making an object, machining creates chips and swarfs that are waste. AM does not create chips because it does not make an object by machining. There may be a small component of machining in AM in the form of modification or post-processing that creates chips—an amount miniscule in comparison to that created in machining. AM is better than machining because it creates far less amount of chips. There is no reason to doubt about the supremacy of AM over machining if creation of chips is measured.

But, there is a reason to doubt about this method for creating supremacy. AM does not intend to do machining and, therefore, if AM is free from the demerit of the machining in comparison to a process that does only machining, then the comparison is not on equal footing. The comparison would have been on equal footing if AM were doing as much machining as a process that does only machining—it could have been great if even then chips created by AM were miniscule.

It does not imply—if a process makes a part without machining (or without creating chips similar to machining) is not an achievement, or creation of chips should not be discouraged, or the component of machining in AM should be increased in order to have a fair method for comparison. But it implies that the method for comparison is undue because it hides the fact that though AM does not make more chips, it may make more waste, or waste energy [18].

For example, if a nylon pattern is made by PBF, it will create waste by degradation of polymer powders present in a bed. The degradation due to PBF will be more than the waste due to the formation of chips if the pattern is instead machined from a block. Using PBF will need more energy than that required for running a CNC machine, and thus the fabrication by PBF will waste more energy. This example does not represent all examples in AM but shows that AM is not always better than machining.

7.3   Material Properties

Machining does not change mechanical properties except some possible changes on a surface due to tool-induced heat and, therefore, properties of material before and after machining remain the same. AM does not come with such material blocks and their properties. It comes with feedstock, experimental parameters, and noise—these all have influences on properties. Machining can be interesting because material properties will not deteriorate due to the lack of skill in machining operation, or it can not be interesting because material properties will not improve even by adopting the best practice.

AM can be interesting because properties at various locations of a part can be changed by optimizing parameters to fulfill different functional requirements—this advantage is not available in machining [19]. AM can not be interesting because material property is an aggregate of contributions provided by particles, pores, voxels, or volumes and, if any entity is not controlled, the property may vary—machining is free from this challenge [20]. But, this challenge has given rise to opportunities—if addition of tiny volumes of materials is controlled during manufacturing, the properties can be predicted—the properties can be known without measuring them. Then the desired property can be introduced at a design stage [21].

8   Disadvantages of ALM

Conversion of a CAD model into layers has an advantage—the problem of making a complex 3D part becomes a problem of making complex 2D layers. Since the fabrication of a complex 2D layer is easier than the fabrication of a complex 3D part, ALM replaces the difficult fabrication process by a simpler one. However, ALM has disadvantages as follows.

8.1   Staircase Effect

During fabrication of a curved part, layers do not exactly coincide with the periphery of a curve, which causes gaps between layers and the curve. These gaps are inherent deficiency of ALM and cannot be eliminated because a straight line will never coincide with a curve.

If a planar layer is replaced with a curved layer, gaps can be eliminated in some geometry. But, there is no AM process that is developed for curved layers, though there has been attempts with extrusion based deposition and curved laminates either to make parts or to make features on an existing part [22].

The cause of the staircase effect is the progress of a feature making an angle with a vertical direction. For example, if a right cylinder is made with its base on a platform, there will not be any staircase effect because the build direction does not make an angle with the central axis of the cylinder (Fig. 4a). The cylinder is made up of horizontal planar layers with circular peripheries (disc), these discs are vertically aligned constituting a cylinder. There is no gap and, therefore, no staircase effect.

But, if the same cylinder is fabricated in a tilted position, horizontal planar layers or discs are required to fill up the tilted cylinder. Since discs grow in a vertical direction, they have vertical straight boundaries. These boundaries do not conform to the oblique

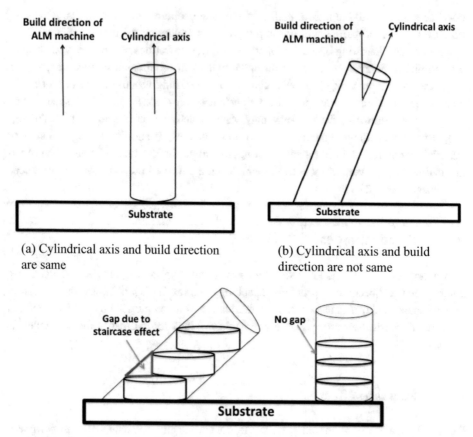

(a) Cylindrical axis and build direction are same

(b) Cylindrical axis and build direction are not same

(c) Gap due to staircase effect in oblique cylinder, no gap in vertical cylinder

Fig. 4 Staircase effects: **a** an example of no staircase effect, **b** an example of staircase effect, **c** gap and no gap

boundary of the tilted cylinder. The gap thus generated between the vertical and oblique boundaries is a staircase effect (Fig. 4b, c). If the discs had oblique boundaries, there would not be any staircase effect.

If the build direction is oblique, the disc will have oblique boundaries. If the build direction is not fixed but tilts with a tilt in a feature with an aim to match the contour of the feature, there will not be any staircase effect. Therefore, if a process has variable build directions, this will be an antidote to the staircase effect.

This effect arises because of the position of one layer over another layer (Fig. 4c, showing gap). This implies if a layer is perfectly made, geometries within the layer are perfectly demarcated and filled up, even then these will have no bearing on the staircase effect.

8.1.1   Removal

To remove the staircase effect, the gap needs to be removed, which can be done in two ways:

(1) It can be filled up. But, filling up is not possible because in the most applications, the gap is too small to be filled up. Even if it can be filled up, this may not be a convenient option as this will take time, which will decrease fabrication speed. Though, it can be filled up in some other applications such as in construction where, for example, a big house is made using AM, and the gap is big enough to justify filling.
(2) The corner encompassing the gap is machined away, if possible. When the corner is gone, there is no gap left. But, this makes the part smaller. In order to have the same part size, the part needs to be made bigger. The increase in the size to get the original part size is called offset.

In Fig. 4c, to remove a gap, a machining tool needs to move parallel to the side surface of the cylinder so the red triangle is removed. This means, after machining, the side surface of the cylinder passes through the tip of the red triangle, which decreases the size of the cross-section of the cylinder. The maximum decrease in the size of the cross-section is equal to the distance from the tip to the base of the triangle or to the side surface. While the minimum decrease in the size is zero.

The cylinder is tilted in an x-direction in an x–y plane. Therefore, when the disc is visualized to be fitted into the cylinder, the maximum gap is in the x-direction. The amount of the gap depends upon the angle of the tilt. If there is more tilt or the angle is big, the gap is more or the staircase effect is more. But, in any case, there is no gap in the y-direction. Therefore, when machining is done to remove the gap, maximum material is removed in the x-direction while no material is removed in the y-direction.

Since the cross-section of the disc is circular, it means the diameter of the disc in the both x- and y-directions is the same. When the material is removed from the x-direction, the diameter of the disc (or cylinder) in the x-direction becomes smaller while no material is removed from the y-direction, the diameter of the cylinder in the y-direction remains unchanged. This causes the circular cross-section of the cylinder to change into an elliptical cross-section. The big diameter of the ellipse is the same as the circular cross-section of the disc while the small diameter is equal to big diameter minus the separation due to the maximum gap caused due to the tilt.

To remove the staircase effect, when the machining is employed, the milling tool has to remove materials from the tilted cylinder. The milling tool has to remove no material in y-direction and maximum material in x-direction when it is moving around the disc of the tilted cylinder. The spinning tool has just to touch the periphery of the disc in the y-direction without removing any material and has to move towards the periphery in the x-direction with an aim to remove maximum materials. The job of the tool is to move from the point of no material removal to the point of the maximum material removal, gradually increasing its removal amount.

When the tool starts to remove the material, it needs to remove the minimum material. How much minimum it can remove depends upon the precision of the tool. How much minimum removal can be accepted depends upon the requirement of an application. If a tool is not able to furnish that requirement, a finer tool is needed. When the tool reaches another point at the periphery where it needs to remove maximum material, how much maximum it can remove depends upon the robustness of the tool. If the tool is not robust, a more robust tool is required.

Thus, the requirement of a tool is to be fine and robust. This requirement may be conflicting if the height, diameter and tilt angle of a cylinder is high. If the diameter and tilt angle is not uniform, i.e. they increase and decrease frequently, the unavailability of the best tool is not a problem but the reach of the tool to all periphery points of the cylinder is the problem. Thus, machining as a means to remove staircase effect is ridden with shortcomings.

The transformation of a circular cylinder, after machining, into an elliptical cylinder can be avoided if original layers are changed. If, instead of original circular layers or discs, bigger size elliptical layers or discs are made, then an elliptical cylinder will be transformed into a circular cylinder after the machining. This can be a solution but this is not an ideal solution as making a bigger part only to have a smaller part by machining does not fit with the goal of less carbon emission. Besides, this solution does not help have an AM that will be free from machining, though it can help machining for AM improve.

8.2   Need for Support Structure

A tilted structure can be self-supporting but if the tilt exceeds a certain angle depending on the weight and geometry, a support structure is required to maintain ongoing fabrication and to counteract potential collapse [23]. The support structure increases the time of fabrication and post-processing, and requires extra material and cost [24].

8.3   Problem in Repair

If the feature of an AM part is broken, then to repair it layerwise, the fabrication steps need to be revisited [25]. If the broken feature is confined within the last few layers, the part is machined till the feature is removed. The removal of the feature may not be possible unless other nearby features are simultaneously removed.

If the removal of a feature is possible without removing other nearby features, this does not necessarily imply that the deposition of a repairing material is possible without getting hindered by nearby features. This may warrant the removal of both broken and nearby features. Thus, the nearby features do not need repair but can be removed.

Depending upon the geometry, there might be several unbroken features that need to be sacrificed to repair a single broken feature. For example, considering the case of a spur gear, which is fabricated in an AM machine mostly in an horizontal orientation, resulting all teeth of the gear to rest on a substrate. If the gear gets damaged during service and needs to be repaired, it will be more convenient if it gets repaired in the horizontal orientation.

Since a tooth extends from the first layer to the last layer of the part (in the horizontal orientation), machining to remove the broken tooth will cause removal of all layers. Thus, removal of one tooth means removal of all teeth. In this case, repair is not possible as there is no machined part remained to be repaired upon, any attempt more to repair the part will be akin to its replacement. Though, it can be repaired by changing the orientation after capturing the image of the damaged part by reverse engineering [26].

For example, it can be repaired in a vertical orientation, in which few teeth rests on the substrate, implying the damaged tooth can be adjusted to be positioned at the top end. In order to repair it, the damaged tooth can be milled off, which means few top layers of the gear will be removed. The repair can then be done by adding some layers on the milled surface.

The example of the horizontal orientation is shown to demonstrate that the benefit of AM achieved in the repair is no longer achievable if the damage is not confined within the last few layers. It shows that the damage done at a certain location of a part is not repairable because attempt is done to repair layerwise. Thus there is a flaw in ALM, which is visible when the attempt is done to repair.

In an ideal case, a repair needs to be confined to a damaged site so the repair can be accomplished without machining and repairing larger area than the area of the damaged site. It is possible if an AM process exists that does not see manufacturing through the prism of a layer, and secondly has the ability to have direct access to the problem site. In ALM, a tool (energy beam, nozzle) is exposed only to the upper side of a layer. Thus, the tool has no access to either lower side or edge (side surface) of the same layer.

There is no need for the tool to be concerned with other sides other than just the upper side because ALM happens only through the upper side—it is not an exaggeration if ALM will instead be called 'upper side ALM'. Since there is no need for the tool to work on other sides, the tool does not work on other sides. What if there is a need—a damaged part is kept on a platform and there are damages at various locations of the part, and damages can be repaired if the tool can manoeuvre and access them.

Even if there is a need as such, the tool can not directly approach the damaged sites, the tool will only move through the layers, the process does not give freedom to the tool to move arbitrarily—if layerwise has given advantages, these have come at a cost.

There are AM machines that are equipped with tools fitted in multi-axis setup or multi-axis robot, but this is the capability of the machine [27] and not that of a process. The tool equipped in such machines is still not free from limitations imposed by the process when the machine uses the tool to perform the process.

9   Additive Non-layer Manufacturing

There are ANLM such as CNC accumulation [28], CLIP [29], 2PP [30], etc. Solvent-cast direct writing technique is used to make a spiral structure, which is non-layerwise fabrication [31].

These processes are emerging and have not found the same acceptance as ALM. Though, they provide a different methodology, which could be of interest if the methodology helps emerge new AM processes to make other than polymer parts. Since ANLM are free from layers, they are also free from the demerits that arise due to layerwise. Some of ANLM, which are photopolymer based, are given below.

9.1   CNC Accumulation

In this process, a substrate is immersed in photopolymer, and a part is fabricated inside the liquid (Fig. 5). To start curing liquid from the substrate, light beam needs to reach the substrate without solidifying the liquid on the way. It is possible if the beam is guided through an optical fibre. The beam then will solidify the liquid between the fibre tip and the substrate, which gets attached on the substrate and the fibre tip. Teflon film on the tip is used to decrease the adhesion between the solidified material and the tip causing the material to attach solely on the substrate. An increment in the attached material happens as per the requirement rather than as per the pre-set layerwise values, thus making the process ANLM.

The attached material grows in the direction of the tip. When the tip moves, the solidified material will follow it, i.e. controlling the movement will create a desired structure.

Fig. 5 Schematic diagram of CNC accumulation

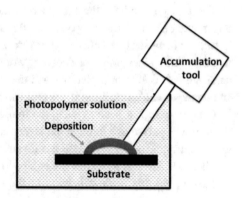

For the tip to be able to move, it needs to be stiff. The tip is supported inside a plastic rigid body and is called a tool. Movement of the tool creates a structure and can repair parts. The movement of this tool (henceforth, this movement) is different from the movement of a tool in bed process (henceforth, that movement).

While creating a structure, this movement unlike that movement is not confined to a layer or a plane. This movement unlike that movement furnishes many build directions and, therefore, the process is relatively free from staircase effect. This movement creates structures on many sides of a substrate, at least three sides of a rectangular substrate, which gives the process an ability to repair a part having damages on many sides without changing the orientation. While that movement creates structures only on an upper side of a substrate. This ability gives the process an edge over bed process [32]. Though, this ability does not give an edge over deposition process as these processes can facilitate multi-orientation deposition [33, 34].

The movement of the tool is facilitated by attaching it on a CNC machine, which is meant to add materials. A tool comprising of a large beam diameter and high laser power can add more material and speed up the fabrication while a tool comprising of a small beam diameter will add small amount, causing an increase in resolution [35]. A combination of tools can thus enable the fabrication of various geometries.

9.2   Continuous Liquid Interface Production

In inverse stereolithography, where a part is made upside down, a film is attached on the inside of an exposure window so the solidified layer can be detached from the window before resin flows for the next layer to be formed. In CLIP, oxygen permeable window instead of a film is used (Fig. 6). Oxygen gas entering through the window reacts with free radicals and neutralizes them so that they will not be free for further polymerization on or near the window.

In the absence of polymerization, there will not be any solidified material on the window required to be detached from. This makes the process free from problems of: detachment, repositioning the substrate after detachment, breakage of features during detachment, not making large area because of the problem of detachment. Utilization of oxygen molecules to inhibit photopolymerization makes the process different from stereolithography where inhibition by such method is not used.

CLIP is a continuous process—there is no recurring phenomenon of layer formation, and there is no recurring detachment. Therefore, to make a part continuously without being interrupted layerwise, photopolymer through window is continuously exposed to beams while a substrate is continuously pulled up.

Continuous pulling up the substrate ensures that the photopolymer continuously flows in between the window and the substrate, which is continuously cured by beams. Fast pulling up will not let complete curing, which means there will be non-uniformity or

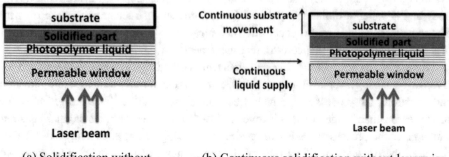

(a) Solidification without   (b) Continuous solidification without layerwise
solidifying on window

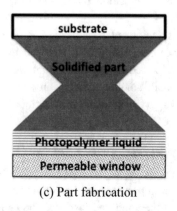

(c) Part fabrication

Fig. 6 Schematic diagram of continuous liquid interface production

porosity in a part. Slow pulling up will cause overcuring, making the part brittle, and decrease the fabrication speed.

Since the process does not proceed layerwise, the process saves time that could have been spent during inter-layer processing, i.e. the process saves processing time after completion of one layer and before starting the next layer, which increases the fabrication speed. Since there is no inter-layer joining, there are no seams between layers that could influence mechanical properties or change isotropicity.

However, the process is not free from a layerwise concept. The exposure on photopolymer or on window by beam makes 2D contour of a part in the same way as happens in a layerwise process. For example, if a solid cone is to be made, the exposure on window will be of circular shape in either process (CLIP or layerwise). After some duration of the exposure, the exposure by beam will no longer be of the same size of the circle, it will change to fit the contour of the cone. When it will change, it will again be same to some nth layer of the layerwise process. In layerwise, geometry of the exposure by beam changes after a progress of a minimum of one layer thickness, while the size of layer thickness is determined by the machine.

In CLIP, geometry of exposure by beam can change early or fast because it is not related to an actual layer thickness a machine can allow, and is not limited by the machine. Since build direction does not change in both processes, there should be staircase effect in both processes. If staircase effect is not visible in CLIP, it is due to minimization of equivalent layer thickness in CLIP. This type of minimum layer thickness can not be set in an actual layerwise machine because of the limitation or tolerance of a machine. CLIP thus demonstrates how the limitation of an ALM machine can be overcome. This also shows that staircase effect is though a limitation of a process but this limitation is pronounced not due to the limitation of the process itself but due to that of the machine.

In layerwise, if a 3D solid cone is fabricated, the first layer will be made by scanning or exposing the photopolymer by a beam in a circular area. The effect of the scanning is the formation of a solid cylinder having height equal to the layer thickness, this cylinder instead of a cone is the natural outcome of a layerwise process. Though, the aim is to make a cone but the cone is not made, it is actually approximated by making a number of cylinders, where each cylinder due to one layer.

What if cone upon cone is made to make a cone. The layerwise process does not make a cone because the process does not see a cone during processing, the process only sees a circle and is not allowed to see the extension of a circle in 3D. CLIP is not different, it does not try to look beyond the plane though it is a continuous process, the beam can look beyond the plane, it can penetrate through the plane and solidify as per Beer-Lambert exponential law of absorption [36]. But, CLIP does not use the beam to go along the contour of a cone or to penetrate obliquely to make the cone directly rather than indirectly by approximation. If CLIP uses beam penetration as per the contour of a part extended in 3D, the process will remove staircase effect. CLIP does not use the beam differently than how it is used in other layerwise process. It is no exaggeration if CLIP is considered a very fast layerwise process having no time to make a layer.

9.3 Two-Photon Polymerization

Making a layer can be a problem if 100 nm resolution is required. 2PP (Fig. 7) makes sub-100 μm size parts having sub-100 nm resolution for medical, optical and acoustical applications [37].

L-beam consisted of femtosecond pulse is required to excite photopolymer and solidify it. In order to excite, two photons within an interval of 10^{-15} s is required. One photon excites the molecule to an intermediate virtual state while another photon excites it from the intermediate state to the final state. If another photon is not available before the molecule loses its intermediate state and comes back to the ground state, the molecule will not be able to reach a higher excited state.

Fig. 7 Schematic diagram of
two-photon polymerization

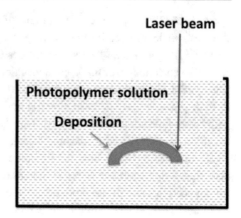

The absorption of two photons present in space and time made available through an
l-beam leads to excitation of polymers that causes polymers to form either free radicals
or cations for polymerization. Energy contributed by two photons is necessary to excite
a molecule. If the same energy is contributed by a single photon, the molecule will not
excite because of the minor difference in quantum states in two cases [38]. Therefore, 2PP
is radically different from single photon polymerization, which is used in photopolymer
bed process.

The probability to find two photons within such a small time is possible only in an
environment of high density of photons, such environment is found at the focal point
of an l-beam. Therefore, when the beam irradiates the liquid, only a small volume of
photopolymer present near the focal point gets cured while the vast amount of the liquid
facing the beam remains unchanged.

This small amount of solidification makes the process slow, and even a sub-mm size
part is too big to be made. But, the same amount gives the process a capability to make
small features and have a high resolution. The smallest feature made in a beam based
process is limited by diffraction limit imposed by the beam, while 2PP is able to make far
smaller features than the wavelength of the laser used, showing the particle characteristic
of light is dominating over its wave characteristic.

Tracing the beam in liquid will leave a trail of solidified material and this is how a
3D feature is made by scanning the beam as per geometry without going through layers.
The features can be made on the surface or inside the liquid. Making inside the liquid
will free the process from availing feedstock (liquid) at reaction sites as the feature made
inside the liquid remains self-supported, which neither sinks nor buoys.

10   Summary of ALM Processes

All major processes along with their most prevalent names are given.

(1) **Additive Friction Stir Deposition (AFSD)**: Powder or rod is deposited and joined on a substrate using friction and stir caused by a tool [39] (Chap. 3).

(2) **Aerosol Jetting (AJ)**: Aerosol is deposited to make micro scale objects [40], also known as aerosol jet printing [41] (Chap. 3).

(3) **Binder Jet Three Dimensional Printing (BJ3DP)**: Binder is jetted on a powder bed (PB) to make layers. Also known as 3D printing [42], binder jetting, binder jetting printing [43] (Chap. 2).

(4) **Ceramic Laser Fusion (CLF)**: Slurry is spread on a platform and is shaped by an l-beam [44], also known as layerwise slurry deposition [45]. In one of its variants named selective laser gelling, sol of the slurry gets converted into gel by an applied l-beam [46, 47]. In another variant, named selective laser gasifying of frozen slurry, ice of a frozen slurry bed is gasified by l-beam to make porous structures [48].

(5) **Cold Spray AM (CSAM)**: Without melting, material is projected at a high speed on a platform [49] (Chap. 3).

(6) **Digital Light Processing**: Selective solidification of photopolymer liquid is done by digital micromirror device [50].

(7) **Electrochemical AM (ECAM)**: Ions are deposited on an electrode [51] (Chap. 3).

(8) **Electron Beam Melting (EBM)**: It is e-beam based PBF, where powders are joined by an e-beam [52] (Chap. 2).

(9) **Fused Deposition Modeling (FDM)**: It is extrusion based AM in which polymer based filaments are melted and extruded. Also known as fused filament fabrication [53] and material extrusion (Chap. 3).

(10) **Fused Pellet Modeling (FPM)**: It is extrusion based deposition in which polymer based pellets are melted and extruded [54]. Also called fused layer modeling [55] and big area AM.

(11) **Fused Deposition of Ceramics (FDC)**: It is extrusion based deposition in which ceramic mixed with binder is extruded [56]. Its other names are aqueous based extrusion fabrication [57], ceramic on demand extrusion [58], robocasting [59, 60], freeze-form extrusion fabrication [61], direct writing (DW) [62], direct ink writing [63] etc.

(12) **High Speed Sintering (HSS)**: Each layer is heated with an infrared lamp after selectively jetting ink on the layer [64, 65]. Its variant with an option to deposit ink at the boundary is multi-jet fusion (MJF) [66] (Chap. 2).

(13) **Ink Jet Printing (IJP)**: Ink is deposited on a platform [67] (Chap. 3). Also known as material jetting, multi jet printing [68].

(14) **Laser Metal Deposition (LMD)**: Powder or wire is joined by an l-beam. It is also called laser direct metal deposition [69]. Also known as LENS when powder is used. Also known as directed energy deposition.

(15) **Laser Engineered Net Shaping (LENS)**: Powders are blown on a platform and joined by an l-beam. Also known as laser powder deposition [70]. In one of its variants when a single crystal is grown, the process is named scanning laser epitaxy [71] (Chap. 3).

(16) **Lithography Based Ceramic Manufacturing**: It is SL using photopolymer based ceramic slurry [72].

(17) **Localized Microwave Heating Based AM (LMHAM)**: Microwave energy supplied by a microwave applicator acts as a heat source to sinter a powder layer [73] (Chap. 2).

(18) **Micro Droplet Deposition Manufacturing (MDDM)**: Droplets due to a low melting point metal are deposited using a nozzle [74].

(19) **Microheater Array Powder Sintering (MAPS)**: This is a variant of SLS in which an array of micro heater replaces l-beam as a heat source [75] (Chap. 2).

(20) **Photopolymer Jetting (PJ)**: Photopolymer is deposited on a platform [76].

(21) **Plasma Arc AM**: Wire is melted by a plasma beam, which is then deposited. Also known as rapid plasma deposition [77] (Chap. 3).

(22) **Powder Melt Extrusion**: It is extrusion based deposition in which polymer powders are melted and extruded [78]. Also known as composite extrusion modeling when composite powders are used [79].

(23) **Rapid Freeze Prototyping (RFP)**: Water drop is frozen to make an object [80]. Its variant is cryogenic prototyping [81] that uses a solution instead of water (Chap. 3).

(24) **Selective Heat Sintering (SHS)**: It is PBF in which fusion is done by thermal printheads [82] (Chap. 2).

(25) **Selective Inhibition Sintering (SIS)**: It is a variant of BJ3DP in which anti-binder ink instead of binder ink is selectively deposited. During post-processing heat treatment, inked area is not joined while remaining area is joined [83] (Chap. 2).

(26) **Selective Laser Melting (SLM)**: Powders are spread on a platform and joined by melting using an l-beam. Also known as direct metal laser sintering. In one of its variants, named microwave assisted SLM, microwave energy is used to preheat a substrate, which removes the requirement of high laser power to melt ceramics [84] (Chap. 2).

(27) **Selective Laser Sintering (SLS)**: Powders are spread on a platform and joined by an l-beam. This is also called laser sintering. In one of its variants, named selective laser flash sintering, electric field is applied across PB to improve sintering rate [85] (Chap. 2).

(28) **Stereolithography (SL)**: Photopolymer is spread on a platform and solidified by a scanning beam. Also known as vat photopolymerization, scan based polymerization, micro-stereolithography and large area maskless photopolymerization [86] (Chap. 2).

(29) **Thermoplastic 3D Printing (T3DP)**: Hard particles mixed with thermoplastics are deposited [87].

(30) **3D Gel Printing (3DGP)**: It is slurry deposition in which slurry is mixed with cross-linking polymers for binding [88] (Chap. 3).

(31) **Wire Arc AM (WAAM)**: Wire is melted by arc, which is then deposited [89] (Chap. 3).

(32) **Wire Electron Beam AM (WEAM)**: Wire is melted by an e-beam, which is then deposited [90] (Chap. 3).

11   Process Differences

AM processes are similar and different in many respects. They are classified earlier [91]. They are different due to: materials, the way materials are joined, the form of feedstocks, and the conveyance of feedstocks. These differences are described below, and attempts are made to classify them as per their differences.

12   Difference Due to Materials

Processes differ from each other on the basis of materials they process. For example, FDM uses polymer based materials, and differs from other processes because it works with polymers. The concept utilized in FDM is not unique to polymers and can be used for metals and ceramics [92]. But, FDM is related to polymer and is not free from it. Similarly, other processes are distinguishable on the basis of a particular material: photopolymer is related to SL, metal is identified with WAAM, etc.

Therefore, a single material can be used to identify one or some processes but not every processes. For example, SLS works with all materials. It can not be identified with a single material (polymer, ceramic, or metal), instead all are required. Hence, a material as an identifying probe is not suitable to identify a process. Thus, a material has a limited capacity to distinguish one process from another.

12.1   Attempt to Classify

If processes are classified on the basis of materials, it will guarantee that all processes get classified as there is no process available that does not deal with a material. For processing a specific material, there are a number of AM processes available. For example, for metals, processes available are LENS, SLS, SLM, EBM, etc. Thus, there are a number of processes that can not be separated from each other because they all process metal (the same material).

If metal is further divided into various groups: iron based alloy, titanium, nickel, etc., there are again a number of processes available that process a particular metal or alloy. It does not imply if the same type of metallic alloy is processed by these processes, they all will not give different properties. Even if they give different properties, it will not lead them to have different places in a classification as the classification is to be on the basis of materials and not of material properties.

In case material properties are included as material plus material property can have a better prospect to become a basis for the classification. Then, a particular process will not have a single place but many different places, because a particular process will process many materials resulting in many specific properties. It will not lead to an ideal goal—a single place for one process.

There is not a single alloy available that should be processed by a dedicated single process while that process should not suppose to process any other metals other than this single alloy. Though, there are customized materials available, which are developed for a specific machine, and are processed by that specific machine. But, a customized material is specific to a machine and not a process. Besides, availability of that customized material does not preclude other customized materials to be developed for that specific machine. There are no specific materials available, each of which corresponds to one process. Therefore, a classification attempted on the basis of materials will not find enough specific materials to distinguish between any two processes.

13   Agents for Joining Materials

Materials are joined by various means: high energy beam (laser, electron, plasma), other thermal source, low temperature, friction energy, binding agent, catalyst, kinetic energy, electrochemical energy, etc. Examples of the processes related to these means or agents are given below:

L-beam—it is used as a thermal source, e.g. SLS, SLM, LENS, CLF, etc. Besides, it is used as a source of photons, e.g. SL, PJ, etc.
E-beam—it is used as a thermal source, e.g. EBM, WEAM.

Plasma beam—it is used as a thermal source. It is obtained from tungsten arc welding equipment.

Other thermal source—micro heater, arc welding used as other non-beam thermal sources, e.g., MAPS, WAAM, etc.

Friction energy—it is used in AFSD, etc.

Kinetic energy—it is used in CSAM.

Binder—it is used in IJP, BJ3DP, etc.

Catalyst—it is used in 3DGP.

Electrochemical energy—it is used in ECAM, electrophoretic deposition (Chap. 3).

Low temperature—not very high temperature provided by a heat source is required e.g. FDM, FPM.

Negative temperature—it is used in RFP, cryogenic prototyping (Chap. 3).

Above list is not exhaustive. Moreover, some processes use more than one means of joining, while the above list gives the one that is responsible for shaping. For example, CLF uses both binder and l-beam, but it is the l-beam that is responsible for shaping. Therefore, CLF is included under l-beam.

13.1 Attempt to Classify

If classification is done on the basis of agents for joining materials, there are some processes that will be well placed in the classification. One example is CSAM, as it is a single example of its corresponding agent, i.e. kinetic energy, there is no need to subclassify as there is no more than one process available that needs to be accommodated. Hence, single agent—single process type case will suit well for CSAM.

However, if an agent has more than one processes, e.g. l-beam, which has many processes such as SLS, LENS, SLM, SL, PJ, etc. [93], then further categories of l-beam need to be searched to find individual places for these processes. If l-beam is categorized as high, medium and low power, the processes will be categorized as high (LENS, SLM), medium (SLS) and low (SL, PJ). This categorization may not be strict because SLS and SLM have overlapping laser power.

Besides, this categorization does not help LENS and SLM to be separated. If l-beam is further sub-categorized as pulse and continuous mode, still both processes (LENS, SLM) can not be separated. There is no more attribute of l-beam available that can separate LENS from SLM. Thus, this classification type has limitations as this is suitable if there is one process and not a number of processes belonging to one agent.

14   Form of Feedstock

A feedstock is a material that is fed into a machine. Following are the forms of the feedstock: powder, wire, liquid, slurry, ion, sheet [94], gas. The form provides a distinctive mark on a process. For example, PBF [95] is distinguishable because it uses a particular form of material, i.e. powder. The importance of the form is more pronounced when it is found that other forms of solid such as wire, sheet, and pellet can not be fitted in this process.

Though, there are processes such as LMD or arc welding based AM, which use more than one form of the solid, i.e. powder and wire. But, using two forms does not trivialize the importance of the form. On the contrary, it emphasizes the uniqueness of the form— the powder is used to furnish accuracy while the wire is used to furnish a high deposition rate. It demonstrates that the selection of the form is governed by the requirement of a part. Thus, there are two processes (LMD-powder and LMD-wire) emanating from one process (LMD) because of a difference in the form.

14.1   Attempt to Classify

If classification is done on the basis of the form, there will be categories: powder, wire, gas, liquid, gel, slurry, etc. To check how this classification will fare, powder is taken as an example. Under the category powder, following processes will come: SLS, HSS, BJ3DP, SHS, SIS, SLM, LENS, CSAM, MDDM, EBM, AFSD, arc welding based processes, etc.

There are many processes under powder, which need to be further classified. A powder has following attributes: small or big, high or low surface roughness, wide or narrow size distribution, flowing or not well-flowing, spherical or non-spherical, and dense or porous. None of these attributes helps these processes to be further categorized as none of these processes works for only a particular attribute. This brings an end to the attempt to classify as there will not be any more sub-categories under powder.

If the form is combined with the material of the feedstock, there can be some possibility to classify. Then, powder can be further divided into following categories: metal, polymer, ceramic, etc. If one of these categories, i.e. metal powder is selected, there will still be many processes: SLS, SLM, LENS, CSAM, MDDM, EBM, AFSD, arc welding based processes, etc. that will come under it. There are no more categories under the metal powder, which will help these processes to be separated. Thus, combining material types with form types still does not help classify.

Keeping these processes together without further classifying will also not help have a classification that gives adequate information. For example, if SLM and LENS are not separated, the classification will not be able to give information about other important aspects of these processes which make them far more different than being similar by dint of them using the metal powder. These aspects are: they are deposited differently, make different products, and do not convert same design into products having the same complexity. Their differences show they must not be placed together.

15   Conveyance of Feedstock

The method of transportation of the same form of feedstock to the point of processing creates a difference between two processes. For example, powder is a form of feedstock for two different processes, i.e. PBF and LMD. One example of PBF and LMD is SLM and LENS, respectively.

15.1   SLM and LENS

In SLM, powder is brought onto a substrate [96] by rolling while in LENS, it is brought by blowing via a nozzle. These two processes are different because of a difference in the conveyance of a feedstock. This difference gives rise to a difference in the timing of the processing.

In SLM, after the conveyance of the powder is over, they are processed by a beam. Their conveyance results in the formation of a powder bed (PB), where they remain immobile before processing happens. The timing of the processing follows their conveyance.

In LENS, powders are conveyed and processed simultaneously by a beam. It means powders are blown, and at the same time beam remains on. The beam makes a melt pool on the substrate, the formation of which is required to receive powders blown into it. The melt pool then causes the powder to melt and become its part. When blowing of the powder and irradiation by the beam happen at the same time, some initial blown powder is lost as the melt pool is not yet created by the beam to receive the powder. But, after the initial loss, powders getting blown is not lost because there is a continuous creation of the melt pool by the beam to receive them. Though, powders will still be lost but the cause of the loss is not the absence of the melt pool as happened during the initial loss.

The cause of the initial loss can be nullified if there will be a time lag between the blowing of the powder and irradiation by the beam. When the beam reaches earlier than the powder, powders will not be lost because there is already a melt pool to receive them. If the beam instead reaches later, more powder will be lost. Thus, the timing difference between the arrival of beam and powder affects the initial conditions of the processing. After the initial condition is over, there is a continuity in bead formation which is due to the immediate conversion of powder into a melt pool.

Thus, in LENS, the initial processing may not exactly coincide with the blowing, but the timing of the processing does not follow the blowing. The processing does not wait for the blowing to be over before it starts.

In SLM, there is a cyclic sequence—powders are conveyed then processed then conveyed and so on. In SLM, there is no processing when they are conveyed or moved. When they are not conveyed, only then are they processed. In LENS, there is no cyclic sequence.

15.2   Other Examples

SL and PJ both use the same form of feedstock, i.e. photopolymer, but these processes differ because photopolymer is conveyed differently. In SL, it is coated while in PJ, it is jetted. In SL, it is cured after it is coated. In PJ, it is cured after it is jetted, but it needs to be cured immediately after it is jetted because delay in curing may lead to flowing or displacement or shape deformation of jetted drops. In SL, there is no such immediacy, there are no jetted drops that are about to leave their positions unless confined by curing, a coated layer can wait longer before getting cured.

The difference between SL and PJ is not only in the transportation of feedstock but also in the timing of their processing (curing) after the transportation. In SL, processing happens when feedstock does not move while in PJ, processing happens when it has just stopped moving. There is a difference between SL and PJ in the timing of processing in relation to their movement.

CLF and T3DP both use slurry as a feedstock, the difference between CLF and T3DP is the difference in the conveyance of the feedstock. In CLF, the slurry is coated while in T3DP, it is deposited. In CLF, a coated slurry is processed by an l-beam. In T3DP, deposited slurry is processed by drying. This drying is not the only processing that is performed in this process, but, this processing is sufficient enough to allow next layer of slurry to be deposited. The processing is expected to be fast lest the deposited slurry be deformed; it needs to be faster if the slurry is of lower viscosity. In T3DP, there is an urgency in drying. In CLF, there is no such urgency in processing, the coated slurry can remain unprocessed for longer duration.

There is a difference in CLF and T3DP for conveying the feedstock. This leads to a difference in the timing of processing after conveying. This difference in timing is observed in SL and PJ, and in SLM and LENS.

There is a relation between CLF and T3DP in terms of the similarity of feedstock material, the conveyance of feedstock, and the difference created. In these terms, there is also a relation between SL and PJ, and SLM and LENS. Thus, there is a relation each between CLF and T3DP, SL and PJ, and SLM and LENS. But these relations tend to obfuscate stronger relations existing among them.

These suggest that CLF, SL, SLM can be bracketed together because the common operation among them is that there is a substantial delay in the processing after the conveyance of feedstock. On the basis of this common operation, there exists a relation among them. While T3DP, PJ, LENS can be bracketed together because there is no such substantial delay in the processing after the conveyance of feedstock. On the basis of this commonality, there exists a relation among these processes as well.

There are two new relations. Thus, all six AM processes can be divided into two major categories, each category for each bracket or each new relation. The division of six processes into two major categories suggests a method to divide all AM processes into two major categories.

15.3 Attempt to Classify

There are many ways a feedstock is conveyed: coating (SL, SLS, EBM, SLM, BJ3DP, etc.), blowing (LENS, CSAM, powder based arc welding process, etc.), powder feeding (AFSD), rod feeding (AFSD), material jetting (IJP, PJ, MDDM, etc.), air jetting (AJ), extruding (FDM, FPM, FDC, etc.), wire feeding (LMD-wire, WAAM, etc.), no feeding (2PP, CLIP, etc.), etc. From the perspective of the conveyance of feedstock, AM processes can be classified into two categories: no feeding and feeding.

15.3.1 No Feeding Category

The classification based on the conveyance of feedstocks needs to take into account those processes where such conveyance does not occur. It is because the conveyance is not required. Materials are not required to be fed because they are already there. For example, in 2PP and CLIP, photopolymer in a container is already present to carry out the process. These processes are not waiting for photopolymer to be either coated or jetted for the next task to perform. Feeding (the photopolymer) is not a component of these processes.

It does not imply that the container will not be fed with the photopolymer in the beginning, or at the end. Or the container will not be periodically replenished. But this type of feeding of the container or filling up the container is not the same as the feeding that happens to be a periodic or a recurring step in a process, as this feeding is not used to bring a predetermined amount of materials, at a certain time, at a certain point of space in a certain fashion without which the process does not proceed.

In all these processes, there is no movement of feeding materials, materials remain still or motionless. These processes can be combinedly called motionless material process. 'Motionless' implies that there is no motion of materials in order to fulfil the feeding, it does not imply there will be no motion due to other reasons such as solidification, phase transformation, depletion of the liquid, environmental agitation, etc.

15.3.2 Feeding Category

There are a few processes that do not require feeding, but the majority of them require. Those processes that require can be divided into two major categories:

(1) Coating
(2) Blowing, feeding, jetting, extruding, spraying.

In the first category, there is a delay between the feeding and its transformation while in the second, there is no such delay. In the first, the transformation starts after the coating of a material is done. In the second, the transformation does not wait for the blowing of powder to be over, i.e. it concurs with the blowing or the feeding of wire, or it does not wait so long when material is jetted or extruded. While in the first, it waits while the material is coated.

Waiting in the first category or lack of such in the second does not connote to the inertia of respective machines. These waitings or no waitings are the inherent characteristics of processes. The waiting emphasizes a sequence in steps.

16   First Feeding Category

In this category, materials are placed on a substrate. For example, in SLS, SLM and EBM, materials in the form of powders are placed; placing the powders is akin to creation of PB. These processes are named PBF. In BJ3DP, PB is formed while in CLF, a slurry bed is formed. In SL, photopolymer is coated, which is akin to creating a photopolymer bed.

In all AM processes related to the coating of feedstock, a material bed is formed. This material bed is either from powder, slurry, or photopolymer. There can be other types of material bed [97]. Processes belonging to this category can be combinedly called material bed process. Since the formation of bed is not possible without the help of material, the use of 'material' in the material bed process is not a necessity and can be dropped. The material bed process thus can be called bed process, which can be classified into solid and liquid bed process (Fig. 3).

Solid bed process that is usually PBP can be further classified into PBF and PB non-fusion (Fig. 4). PBF can be classified into complete or partial fusion based processes, depending upon the complete or partial melting of powders, respectively. Melting is induced either by an l-beam or e-beam. Thus complete fusion type processes can be classified into l-beam and e-beam based processes. SLM and EBM belong to complete fusion type induced by l-beam and e-beam, respectively while SLS and HSS belong to partial fusion type. BJ3DP belongs to non-fusion type. This classification can be done in another way as well (Chap. 2).

Liquid bed processes can be classified into photopolymer and slurry bed processes. Photopolymer bed processes can be further classified into scan and projection based processes (Fig. 5). Photopolymer bed process is known as SL (Chap. 2).

Complete classification of the bed process by combining Figs. 8, 9 and 10 is given in Fig. 11.

Fig. 8 Classification of bed
process into two major types

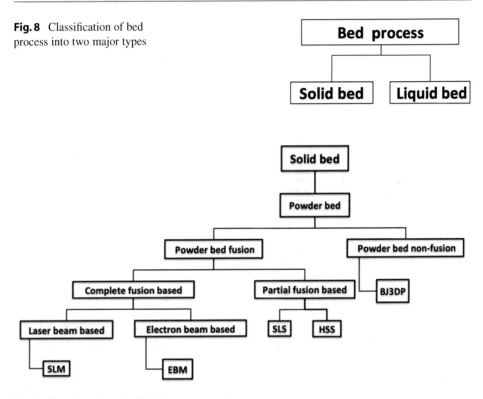

Fig. 9 Classification of solid bed process

Fig. 10 Classification of
liquid bed process

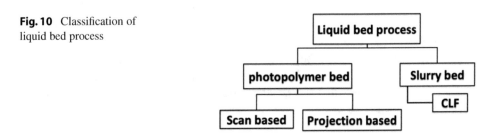

17   Second Feeding Category

It is related to the following types: blowing, wire feeding, jetting, extruding. In all types, a material moves from one point to another while making a structure. In blowing, powder is moved from a nozzle to a substrate. In wire feeding, wire moves from a feeder to a substrate. In jetting, liquid moves from a printhead. In extruding, polymer moves from a nozzle.

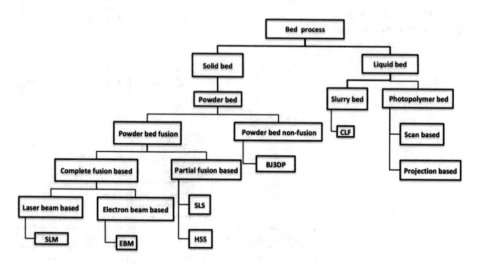

Fig. 11 Classification of bed process

This is different from the first feeding category where materials move from one point to another to make a bed. Their movements do not make any structure, instead they always make the same type of bed. While in this category, materials are moved or deposited to make a design. This is common among all processes in this category. Therefore, all processes can be combinedly called material deposition process.

Material deposition process can also be called deposition process because if there is a deposition, it will be the deposition of the material. There is no other possibility. Thus, the name 'deposition process' will not able to imply any process other than that implied by the material deposition process.

The deposition process can be classified into: solid deposition process (SD), liquid deposition process (LD), air deposition process, and ion deposition processes (Fig. 12).

SD can be further classified into powder, wire [98], filament, rod and pellet (Fig. 13). Powder can be deposited using sources of energy: beam (l-beam, plasma), arc welding,

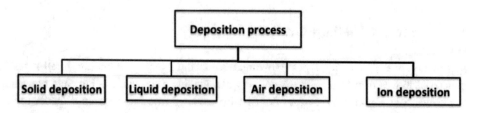

Fig. 12 Classification of deposition process

friction, and cold spray [99], giving rise to the classification of powder deposition processes into processes based on these sources of energy. Powder deposition processes are LENS (based on l-beam), AFSD (based on friction), and CSAM (based on cold spray).

Wire can be deposited using sources of energy such as beam (l-beam, e-beam, plasma), and arc welding, giving rise to the classification of wire deposition processes into processes based on l-beam, e-beam, plasma, and arc welding. Wire deposition processes are LMD (based on l-beam), WEAM (based on e-beam), and WAAM (based on arc welding). Pellet deposition process is FPM or fused layer modeling or big area AM while filament deposition process is FDM.

LD can be classified into processes based on deposition due to polymer, ink (IJP), photopolymer (PJ), metal, water (RFP), and slurry (Fig. 14). Slurry deposition processes can be classified into processes based on photopolymer, polymer (T3DP), and gel (3DGP) (Fig. 15). Air deposition process is aerosol jetting (AJ) while ion deposition process is electrochemical AM (ECAM).

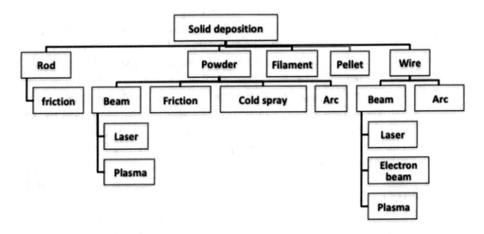

Fig. 13 Classification of solid deposition

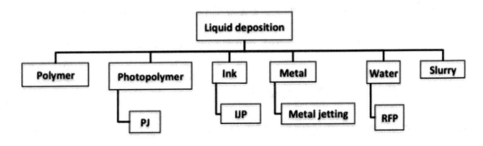

Fig. 14 Classification of liquid deposition

Fig. 15 Classification of
slurry deposition

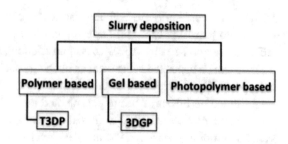

18   Difference Between Solid and Liquid Deposition

SD uses a solid feedstock while LD uses a liquid feedstock. For example, LENS, which
is SD, uses powder as a solid feedstock while IJP, which is LD, uses ink as a liquid
feedstock. Since most of the SD makes products by changing from solid to liquid to
solid, SD requires the control of the flow of the solid feedstock and the liquid that forms.
In LD, only the control of the liquid is required as there is no solid.

What if LD uses a solid feedstock as a source for a liquid feedstock. For example, in
metal jetting [100], it is not convenient for LD system to store liquid feedstock or molten
metal because the system needs to have a storage for it, which needs to be maintained at
a high temperature even if there is no deposition. It is rather more convenient to store a
solid feedstock such as metal powders or rods and melt them to get the liquid feedstock
as and when required. This process will still come under LD, though it uses the solid
feedstock as it is controlled by controlling the amount and frequency of a liquid jet. Solid
feedstock has no direct connection with the deposition and controlling its amount will not
change the amount and frequency of the liquid jet. If it changes, it is what is done in SD,
it is then not LD. The role of the solid feedstock, in LD, is only to ensure that there is a
sufficient amount of the liquid feedstock available to be further controlled for deposition.

In SD, the role of a solid feedstock is to provide a means to directly control the size
and shape of a deposited material. For example, in WEAM, by controlling solid feedstock
rate (wire feed rate), the size and shape of a molten pool can be controlled. In induction
based wire feed AM, it can be controlled by controlling the metallic extrudate [101, 102].
In FDM, the role of a filament (solid feedstock) is to provide a means to directly control
the size and shape of an extrudate.

This is how LD with a solid feedstock differs from SD with a solid feedstock.

19   Classification Due to Feeding

On the basis of feeding categories, AM processes can be classified into three types
(Fig. 16):

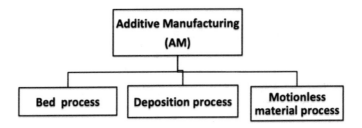

Fig. 16 Classification on the basis of feeding

(i) Bed process (from the first feeding category)
(ii) Deposition process (from the second feeding category)
(iii) Motionless material process (from no feeding category).

Further classification of the first category is given in Fig. 11 while that of the second category is given in Figs. 12, 13, 14 and 15.

20 Combined Classification

Figure 2 states that AM can be classified into ALM and ANLM while Fig. 16 states that it can be classified into three types. It brings a question—whether and why these two classifications need to be merged.

On the basis of the layerwise fabrication, AM can be seen as is going through making or not making layers. But, it does not tell how they are made—it is oblivious of the fact that there are feedstocks involved.

As per Fig. 16, AM can be seen in relation to feedstocks. It conveys how and whether materials are moving or getting deposited but is silent on the consequence of these actions in terms of the formation of layers. The consequence may be the formation of layers or non-layers, but Fig. 16 does not give those information. Combining Fig. 16 with Fig. 2 will give more information.

Figure 17 is a result of this combination in which each type given in Fig. 16 is divided into two types, e.g. bed process is of two types ALM and ANLM, and so on.

Material bed formation itself implies that a layer in the form of bed is formed and is of ALM type. All bed process types shown in Fig. 11 are ALM, implying the impossibility of the bed process as ANLM. Though, there is some possibility to utilize bed type mechanism in ANLM (Chap. 2).

Deposition process shown in Figs. 12, 13, 14 and 15 are of ALM type while there is some processes that are ANLM [103]. Since the deposition process under ANLM are not many, they can be ignored.

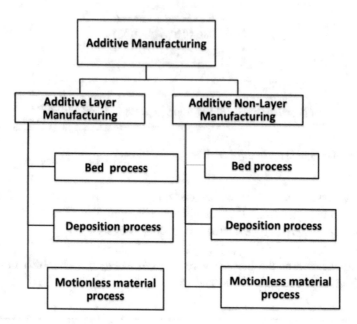

Fig. 17 Combined classification

Most of the ANLM such as 2PP, CLIP come under motionless material process. There is no motionless material type of ALM type, therefore, it is removed from the classification. The final classification (Fig. 18) accommodates all existing AM processes and has scope to accommodate future processes.

Fig. 18 Final classification

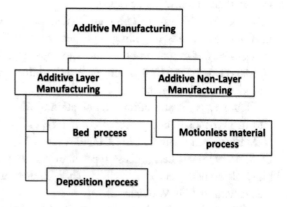

20.1   Was the Decision to Combine Right

Final classification (Fig. 18) is not much different from the classification due to feeding (Fig. 16). The notable difference is that motionless material process is categorized as ANLM. This categorization again does not help for practical purposes as there are few ANLM, and therefore taking into the account whether a process is ALM or ANLM does not make substantial difference in the final classification. This brings a question mark on the categorization of AM into ALM and ANLM, and then its inclusion in the classification.

As final classification (Fig. 18) shows it almost does not include ANLM. The classification is right on the ground that classification is meant to classify something, and if that something is absent, the classification must not include it. The classification of processes means there should exist some process to be included in order to be classified, and not that the process should be made up later in order to validate the classification. This justifies why the final classification does not almost include ANLM.

But, it was known in the beginning there are not many ANLM, then why the effort was made to include in the classification. It is because whatever process is known must demonstrate its classifiability, and therefore it was included even it was meant to be excluded eventually. Secondly, the attempt to include demonstrates where (in the classification) the future processes have scope to be included, if they are warranted.

20.2   Slurry in the Classification

A slurry can be classified independently without being classified as liquid since the slurry is a mixture of solid and liquid. In case the slurry needs to selected as either solid or liquid, it is more suitable as liquid than solid. Because, depending upon the solid content in it, its behaviour can change from low-viscous to high-viscous liquid. Though, the flow behaviour of some slurry resembles to that of the powder (a solid) and, therefore, some slurry gives an impression that it is suitable to be classified as a solid along with powder, but the majority of them do not resemble to the powder.

Some slurry, e.g. frozen slurry does not behave as powder or liquid and is not fit to be classified as liquid. It can be classified as solid if the classification has only solid and liquid category. Therefore, some slurry will be classified as solid while some as liquid. This defeats the purpose to find a single and unique place for the whole slurry (without dividing it) in the classification. The division of the slurry makes the classification bigger as well. It would be better if the classification was succinct without sacrificing the information. If there is no information to sacrifice, it would be easier to classify.

If there is more information, it requires more effort to classify, while for less information, it requires less effort. For example, if there are two or three AM processes, two or three materials processed, two or three products formed, there is no need to classify.

Classification is required because there are many processes, materials and products, which need to be sorted and arranged. Thus, this is the information that leads to the classification. There is information because there is a source for it. The source is what is practised in AM.

What is practised more in AM is more vital information. Again, what is not practised in AM is not a piece of information for the classification purpose. Thus, what is not practised in AM but practised in other manufacturing processes is vital information for some other purposes but is not valuable for the classification of AM. Similarly, what is practised less in AM is less vital information.

In an ideal case, there is no information needed to be sacrificed and the classification should be complete. But if the classification demands some information to be sacrificed, it is the less vital information that needs to be sacrificed to make the classification as complete as possible.

The type of slurry that is similar to solid than to liquid is in minority and can be sacrificed. This makes the whole slurry to be classified as liquid. What will happen if what is practised in AM changes—the slurry that is liquid type becomes obsolete and the slurry that is solid type is the only one used in AM? Then, the slurry will be classified as solid.

As per classification, identity of the slurry changes from liquid to solid with a change in the amount of types of the slurry processed by AM. This identity change of the slurry has nothing to do with how a slurry is formed or what is the chemistry of the slurry changing, or how its behaviour changes from liquid to solid with the passage of time.

The identity change shows how perception about the slurry can change. This demonstrates how the perception about something (slurry) can be made, if it has no separate place in the classification. On the other way round, it shows how a skewed classification can help create an artificial perception.

This shows how the composition of the classification relies on the information fed to it. If the information about something dominates, it will be reflected in the classification. The classification shows what is happening and not the vice versa—it implies that the classification does not decide what should happen. Though, it can be a means to predict the future [97].

21   About the Book

Figure 18 shows AM processes can be divided into three categories: bed process, deposition process, and motionless material process, irrespective of whether they are ALM or ANLM. Therefore, if a book is written about AM process and it is decided that the book will have three chapters, the three chapters will bear the name of these three categories, and the book will not be incomplete. Since majority of AM processes do not fall outside the first two categories, the number of chapters can still be reduced.

The present book contains two chapters representing two major categories, and including the present chapter, there are three chapters: (1) process and classification, (2) bed process, and (3) deposition process.

Confinement of all AM processes within three chapters shows that even if processes are diverse, they can be arranged within a few chapters. The arrangement shows that the classification shown in Fig. 18 is not only for the sake of classification, but it has practical purposes. The present book is the practical manifestation of the classification.

It does not mean that, even if a voluminous book on AM process is written, it should have no more than few chapters to show the practical manifestation of the classification. It also does not mean, in an extreme case, if the whole content of the book is confined within a single chapter, the book will show why a classification is not required and has never been an issue in AM. It only means, for the sake of convenience, there could be any number of chapters, but the arrangement of contents in the present book within a few chapters as per the classification demonstrates that the classification provides a way that is implementable.

What is the implication of this manifestation of the classification? It brings ultimate simplification. It conveys that the world of AM is binary—either bed process or deposition process. It does not mean there is no room for a complex process in AM. It only means there is no room for any complex process in AM, which is free from either bed process or deposition process.

The present book, though it deals with AM process, does not have a title of AM process but of AM classification. What is the reason for the paradox? This is because the book does not deal with all aspects of a process but with mostly those aspects that help justify the composition of the classification. If a book on process classification is written, the book will not be free from the description of the process, which the present book is. The main aim is not to convey that this is how a process should be described, but how the process should be arranged. This does not mean the classification of a process is more important than the description of a process, but it implies that the description of a process is not important unless the position of the process with respect to the positions of other processes is known. This justifies why the book is deprived of the title, the subject of which it is describing.

The book was originally published as 'Additive Manufacturing Processes' [104].

References

1. Button ST (2014) Introduction to advanced forming technologies. In: Comprehensive materials processing, vol 3. Elsevier, pp 1–5
2. Zakeri S, Vippola M, Levänen E (2020) A comprehensive review of the photopolymerization of ceramic resins used in stereolithography. Addit Manuf 35:101177
3. Pan Z, Wang Y, Huang H et al (2015) Recent development on preparation of ceramic inks in ink-jet printing. Ceram Int 41(10):12515–12528

4. Yan C, Shi Y, Zhaoqing L et al (2020) Selective laser sintering additive manufacturing technology. Academic Press

5. Galati M, Iuliano L (2018) A literature review of powder-based electron beam melting focusing on numerical simulations. Addit Manuf 19:1–20

6. Zhang Y, Guo Y, Chen Y et al (2020) Microstructure and mechanical properties of Al-12Si alloys fabricated by ultrasonic-assisted laser metal deposition. Materials 13(1):126

7. Kumar S (2020) Introduction. In: Additive manufacturing processes. Springer, Cham, pp 1–19

8. Kumar S (2022) Synonym. In: Additive manufacturing solutions. Springer, Cham, pp 1–6

9. Yu HZ, Jones ME, Brady GW et al (2018) Non-beam-based metal additive manufacturing enabled by additive friction stir deposition. Script Mater 153:122–130

10. Kumar S (2022) Role of post-process. In: Additive manufacturing solutions. Springer, Cham, pp 41–56

11. Kumar S (2022) Comparison. In: Additive manufacturing solutions. Springer, Cham, pp 57–92

12. Kumar S (2020) Laser powder bed fusion. In: Additive manufacturing processes. Springer, Cham, pp 41–63

13. Roschli A, Gaul KT, Boulger AM et al (2019) Designing for big area additive manufacturing. Addit Manuf 25:275–285

14. Zhou X, Hou Y, Lin J (2015) A review on the processing accuracy of two photon polymerization. AIP Adv 5:030701

15. Kumar S (2022) Fabrication strategy. In: Additive manufacturing solutions. Springer, Cham, pp 111–144

16. Hafkamp T, Baars GV, Jager BD, Etman P (2017) A trade-off analysis of recoating methods for vat photopolymerization of ceramics. In: SFF proceedings, vol 28, pp 687–711

17. Santoliquido O, Colombo P, Ortona A (2019) Additive manufacturing of ceramic components by digital light processing: a comparison between the "bottom-up" and the "top-down" approaches. J Eur Ceram Soc 39(6):2140–2148

18. Kumar S (2022) Application. In: Additive manufacturing solutions. Springer, Cham, pp 93–110

19. Kumar S (2022) Advantage. In: Additive manufacturing solutions. Springer, Cham, pp 7–29

20. Kumar S (2022) Disadvantage. In: Additive manufacturing solutions. Springer, Cham, pp 31–40

21. Roach RA, Bishop JE, Johnson K et al (2018) Using additive manufacturing as a pathway to change the qualification paradigm. In: SFF symposium proceedings, pp 3–13

22. McCaw JCS, Urquizo EC (2018) Curved-layer additive manufacturing of non-planar parametric lattice structures. Mater Des 160:949–963

23. Mezzadri F, Bouriakov V, Qian X (2018) Topology optimization of self-supporting support structures for additive manufacturing. Addit Manuf 21:666–682

24. Paul R, Anand S (2015) Optimization of layered manufacturing process for reducing form errors with minimal support structures. J Manuf Syst 36:231–243

25. Wilson JM, Piya C, Shin YC et al (2014) Remanufacturing of turbine blades by laser direct deposition with its energy and environmental impact analysis. J Clean Prod 80:170–178

26. Anwer N, Mathieu L (2016) From reverse engineering to shape engineering in mechanical design. CIRP Ann Manuf Technol 65:165–168

27. Tsao C, Chang H, Liu M et al (2018) Freeform additive manufacturing by vari-directional vari-dimensional material deposition. Rapid Prototyp J 24(2):379–394

28. Chen Y, Zhou C, Lao J (2011) A layerless additive manufacturing process based on CNC accumulation. Rapid Prototyp J 17(3):218–227

29. Janusziewicz R, Tumbleston JR, Quintanilla AL et al (2016) Layerless fabrication with continuous liquid interface production. PNAS 11(42):11703–11708

30. Wu S, Serbin J, Gu M (2006) Two-photon polymerization for three-dimensional micro-fabrication. J Photochem Photobiol A Chem 181:1–11
31. Guo S, Gosselin F, Guerin N et al (2015) Solvent-cast three-dimensional printing of multifunctional microsystems. Small 9(24):4118–4122
32. Simonelli M, Tse YY, Tuck C (2014) Effect of the build orientation on the mechanical properties and fracture modes of SLM Ti–6Al–4V. Mater Sci Eng A 616:1–11
33. Yang Y, Fuh JY, Loh HT, Wong YS (2003) Multi-orientational deposition to minimize support in the layered manufacturing process. J Manuf Syst 22(2):116–129
34. Yuan L, Pan Z, Ding D et al (2021) Fabrication of metallic parts with overhanging structures using the robotic wire arc additive manufacturing. J Manuf Process 63:24–34
35. Pan Y, Zhou C, Chen Y, Partanen J (2014) Multitool and multi-axis computer numerically controlled accumulation for fabricating conformal features on curved surfaces. J Manuf Sci Eng 136(031007):1–14
36. Jacobs FP (1992) Rapid prototyping and manufacturing: fundamentals of stereolithography. Society of Manufacturing Engineers, Dearborn, MI
37. Nguyen AK, Narayan RJ (2017) Two-photon polymerization for biological applications. Mater Today 20(6):314–322
38. Fourkas JT (2016) Fundamentals of two photon fabrication. In: 3D fabrication using two photon polymerization. Elsevier Inc, pp 45–61
39. Gopan V, Wins KLD, Surendran A (2021) Innovative potential of additive friction stir deposition among current laser based metal additive manufacturing processes: a review. CIRP J Manuf Sci Technol 32:228–248
40. Goh GL, Agarwala S, Tan YJ, Yeong WY (2018) A low cost and flexible carbon nanotube pH sensor fabricated using aerosol jet technology for live cell applications. Sens Actuators B Chem 260:227–235
41. Wilkinson NJ, Smith MAA, Kay RW et al (2019) A review of aerosol jet printing—a non-traditional hybrid process for micro-manufacturing. Int J Adv Manuf Technol 1–21
42. Kernan BD, Sachs EM, Oliveira MA, Cima MJ (2007) Three dimensional printing of tungsten carbide-10 wt % cobalt using a cobalt oxide precursor. Int J Refract Met Hard Mater 25:82–94
43. Kunchala P, Kappagantula K (2018) 3D printing high density ceramics using binder jetting with nanoparticle densifiers. Mater Des 155:443–450
44. Tang HH (2002) Direct laser fusing to form ceramic parts. Rapid Prototyp J 8(5):284–289
45. Muhler T, Gomes C, Ascheri M et al (2015) Slurry-based powder beds for selective laser sintering of silicate ceramics. J Ceram Sci Technol 06(02):113–118
46. Liu FH, Liao YS (2010) Fabrication of inner complex ceramic parts by selective laser gelling. J Eur Ceram Soc 30(16):3283–3289
47. Liu FH, Lee RT, Lin WH, Liao YS (2013) Selective laser sintering of bio-metal scaffold. Procedia CIRP 5:83–87
48. Zhang G, Chen H, Zhou H (2017) Additive manufacturing of green ceramic by selective laser gasifying of frozen slurry. J Eur Ceram Soc 37(7):2679–2684
49. Yin S, Lupoi R (2021) Cold spray additive manufacturing. In: Springer tracts in additive manufacturing
50. Salonitis K (2014) Stereolithography. In: Comprehensive materials processing, vol 10. Elsevier, pp 19–67
51. Kamraj A, Lewis S, Sundaram M (2016) Numerical study of localized electrochemical deposition for micro electrochemical additive manufacturing. Procedia CIRP 42:788–792
52. Gong X, Anderson T, Chou K (2014) Review on powder-based electron beam additive manufacturing technology. Manuf Rev 1(2):1–12

53. Brenken B, Barocio E, Favaloro A et al (2018) Fused filament fabrication of fiber-reinforced polymers: a review. Addit Manuf 21:1–16
54. Wang Z, Liu R, Sparks T, Liou F (2016) Large scale deposition system by an industrial robot (I): design of fused pellet modeling system and extrusion process analysis. 3D Print Addit Manuf 3(1):39–47
55. Kumar N, Jain PK, Tandon P, Pandey PM (2018) Investigation on the effects of process parameters in CNC assisted pellet based fused layer modeling process. J Manuf Process 35:428–436
56. Onagoruwa S, Bose S, Bandyopadhyay A (2001) Fused deposition of ceramics (FDC) and composites. In: Proceedings of SFF, Austin, pp 224–231
57. Mason MS, Huang T, Landers RG et al (2009) Aqueous-based extrusion of high solids loading ceramic pastes: process modeling and control. J Mater Process Technol 209:2946–2957
58. Li W, Ghazanfari A, Leu MC, Landers RG (2015) Methods of extrusion on demand for high solids loading ceramic paste in freeform extrusion fabrication. In: Proceedings of SFF symposium, Austin, pp 332–345
59. Stuecker JN, Cesarano JA, Hirschfeld D (2003) Control of the viscous behavior of highly concentrated mullite suspensions for robocasting. J Mater Process Technol 142:318–325
60. Smay JE, Tuttle B, Cesarano J (2008) Robocasting of three-dimensional piezoelectric structures. In: Safari A, Akdoğan EK (eds) Piezoelectric and acoustic materials for transducer applications. Springer, Boston
61. Huang T, Mason MS, Hilmas GE, Leu MC (2006) Freeze-form extrusion fabrication of ceramic parts. Virtual Phys Prototyp 1:93–100
62. Balani SB, Ghaffar SH, Chougan M et al (2021) Processes and materials used for direct writing technologies: a review. Results Eng 11:100257
63. Shahzad A, Lazoglu I (2021) Direct ink writing (DIW) of structural and functional ceramics: recent achievements and future challenges. Compos Part B Eng 225:109249
64. Brown R, Morgan CT, Majweski CE (2018) Not just nylon—improving the range of materials for high speed sintering. In: SFF proceedings, pp 1487–1498
65. Thomas HR, Hopkinson N, Erasenthiran P (2006) High speed sintering—continuing research into a new rapid manufacturing process. In: SFF proceedings, pp 682–691
66. Sillani F, Kleijnen RG, Vetterli M et al (2019) Selective laser sintering and multi jet fusion: process-induced modification of the raw materials and analyses of parts performance. Addit Manuf 27:32–41
67. Stringer J, Derby B (2009) Limits to feature size and resolution in ink-jet printing. J Eur Ceram Soc 29:913–918
68. Hou MM (2018) Overview of additive manufacturing process. In: Zhang J, Jung Y-G (eds) Additive manufacturing. Butterworth-Heinemann, pp 1–38
69. Pinkerton AJ (2010) Laser direct metal deposition: theory and applications in manufacturing and maintenance. In: Advances in laser materials processing. Woodhead Publishing, pp 461–491
70. Vilar R (2014) Laser powder deposition. In: Comprehensive materials processing, vol 10. Elsevier Ltd, pp 163–216
71. Kirka M, Bansal R, Das S (2009) Recent progress on scanning laser epitaxy: a new technique for growing single crystal superalloys. In: SFF proceedings, pp 799–806
72. Schwarzer E, Götz M, Markova D et al (2017) Lithography-based ceramic manufacturing (LCM)—viscosity and cleaning as two quality influencing steps in the process chain of printing green parts. J Eur Ceram Soc 37(16):5329–5338
73. Jerby E, Meir Y, Salzberg A et al (2015) Incremental metal-powder solidification by localized microwave-heating and its potential for additive manufacturing. Addit Manuf 6:53–66

74. Chao Y, Qi L, Xiao Y et al (2012) Manufacturing of micro thin-walled metal parts by micro-droplet deposition. J Mater Process Technol 212(2):484–491

75. Holt N, Horn AV, Montazeri M, Zhou W (2018) Microheater array powder sintering: a novel additive manufacturing process. J Manuf Process 31:536–551

76. Lanceros-Méndez S, Costa CM (2018) Printed batteries: materials, technologies and applications. Wiley

77. Feng Y, Zhan B, He J, Wang K (2018) The double-wire feed and plasma arc additive manufacturing process for deposition in Cr-Ni stainless steel. J Mater Process Technol 259:206–215

78. Boyle BM, Xiong PT, Mensch TE et al (2019) 3D printing using powder melt extrusion. Addit Manuf 29:100811

79. Lieberwirth C, Harder A, Seitz H (2017) Extrusion based additive manufacturing. J Mech Eng Autom 7:79–83

80. Bryant FD, Sui G, Leu MC (2003) A study on effects of process parameters in rapid freeze prototyping. Rapid Prototyp J 9(1):19–23

81. Pham CB, Leong KF, Lim TC, Chian KS (2008) Rapid freeze prototyping technique in bioplotters for tissue scaffold fabrication. Rapid Prototyp J 14(4):246–253

82. Baumers M, Tuck C, Hague R (2015) Selective heat sintering versus laser sintering: comparison of deposition rate, process energy consumption and cost performance. In: SFF proceedings, pp 109–121

83. Khoshnevis B, Zhang J, Fateri M, Xiao Z (2014) Ceramics 3D printing by selective inhibition sintering. In: SFF proceedings, pp 163–169

84. Buls S, Vleugels J, Hooreweder BV (2018) Microwave assisted selective laser melting of technical ceramics. In: SFF proceedings, pp 2349–2357

85. Hagen D, Kovar D, Beaman JJ (2018) Effects of electric field on selective laser sintering of yttria-stabilized zirconia ceramic powder. In: SFF symposium proceedings, pp 909–913

86. Rudraraju A, Das S (2009) Digital date processing strategies for large area maskless photopolymerization. In: SFF symposium proceedings, pp 299–307

87. Scheithauer U, Potschke J, Weingarten S et al (2017) Droplet-based additive manufacturing of hard metal components by thermoplastic 3D printing (T3DP). J Ceram Sci Technol 8(1):155–160

88. Ren X, Shao H, Lin T, Zheng H (2016) 3D gel-printing—an additive manufacturing method for producing complex shaped parts. Mater Des 101:80–87

89. Tabernero I, Paskual A, Alvarez P, Suarez A (2018) Study on arc welding processes for high deposition rate additive manufacturing. Procedia CIRP 68:358–362

90. Fox J, Beuth J (2013) Process mapping of transient melt pool response in wire feed e-beam additive manufacturing of Ti-6Al-4V. In: SFF proceedings, pp 675–683

91. Kumar S (2020) Classification. In: Additive manufacturing processes. Springer, Cham, pp 21–40

92. Rane K, Strano M (2019) A comprehensive review of extrusion-based additive manufacturing processes for rapid production of metallic and ceramic parts. Adv Manuf 7:155–173

93. Schmidt M, Merklein M, Bourell D et al (2017) Laser based additive manufacturing in industry and academia. CIRP Ann 66(2):561–583

94. Kumar S (2020) Sheet based process. In: Additive manufacturing processes. Springer, Cham, pp 171–186

95. Grasso M, Colosimo BM (2017) Process defects and insitu monitoring methods in metal powder bed fusion: a review. Meas Sci Technol 28:044005

96. Snow Z, Martukanitz R, Joshi S (2019) On the development of powder spreadability metrics and feedstock requirements for powder bed fusion additive manufacturing. Addit Manuf 28:78–86

97. Kumar S (2020) Future additive manufacturing processes. In: Additive manufacturing processes. Springer, Cham, pp 187–202

98. Fredriksson C (2019) Sustainability of metal powder additive manufacturing. Procedia Manuf 33:139–144

99. Dass A, Moridi A (2019) State of the art in directed energy deposition: from additive manufacturing to materials design. Coatings 9(418):1–26

100. Simonelli M, Aboulkhair N, Rasa M et al (2019) Towards digital metal additive manufacturing via high-temperature drop-on-demand jetting. Addit Manuf 30:100930

101. Englert L, Klumpp A, Ausländer A et al (2022) Semi-solid wire-feed additive manufacturing of AlSi7Mg by direct induction heating. Addit Manuf Lett 100067

102. Zhang Q, Li H, Han B et al (2022) A distinctive Pb-Sn semi-solid additive manufacturing using wire feeding and extrusion. J Manuf Process 80:247–258

103. Kumar S (2020) Additive non-layer manufacturing. In: Additive manufacturing processes. Springer, Cham, pp 159–170

104. Kumar S (2020) Additive manufacturing processes. Springer, Cham

Bed Process

2

1   Classification

Bed process is of three types: powder, photopolymer, and slurry (Fig. 1), which can be both non-layerwise [1, 2] and layerwise.

2   Powder Bed Process

Powder bed process (PBP) is a generic name for AM processes in which a powder bed (PB) is created and selectively joined to make a part. PB means a thin layer of powders spread on a platform (or a substrate). The thin layer corresponds to the slice of a 3D CAD model of a would-be part. PBP implies thus creation of PB and its consolidation (joining) thereafter.

Variation in PB or types of joining brings variation in PBP. For example, if PB is compacted by applying pressure, it is no longer similar to PB created by just spreading powders using a roller or scraper (or recoater). The compaction creates another PBP. In other examples of variation, if powders are joined by two types: (1) completely melting using a high energy beam, (2) interlocking using binder, then these two types are not the same PBP.

PBP is found to be of two major types: (1) powder bed fusion (PBF), and (2) powder bed non-fusion (Fig. 2). As per Oxford dictionary, fusion is the process of causing a material or object to melt with intense heat so as to join with another.

In PBF, powders are partially or fully melted to join them, e.g. selective laser sintering (SLS), selective laser melting (SLM), electron beam melting (EBM), selective heat sintering (SHS) [3], micro heater array powder sintering (MAPS) [4], high speed sintering (HSS) [5], localized microwave heating based AM (LMHAM) [6].

© The Author(s), under exclusive license to Springer Nature Switzerland AG 2022   43
S. Kumar, *Additive Manufacturing Classification*, Synthesis Lectures on Engineering,
Science, and Technology, https://doi.org/10.1007/978-3-031-14220-8_2

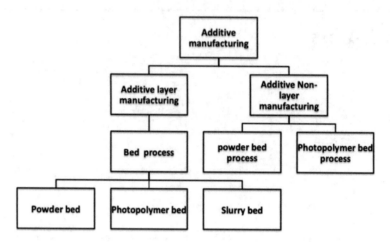

Fig. 1 Classification of bed process

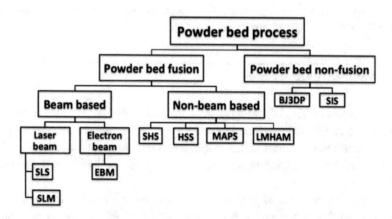

Fig. 2 Classification of powder bed process

In PB non-fusion, powders are not melted, binders are used to join them, e.g. binder jet 3D printing (BJ3DP), selective inhibition sintering (SIS) [7], etc.

PBF can be further divided into two categories: (1) beam based that require high energy beams, e.g. SLS, SLM, EBM, and (2) non-beam based that do not require a high energy beam but works with other thermal sources such as heaters, lamps, microwave, e.g. SHS, HSS, MAPS, LMHAM.

Beam PBF can be further divided into two categories: (1) laser powder bed fusion (LPBF), comprising of SLS and SLM, (2) e-beam powder bed fusion (EPBF), comprising of only EBM.

2.1   Description of Classification

In SLS, if a polymer-coated high melting point material as a powder is used, it is the polymer that is melted, which helps join the high melting point material. Therefore, it is not an example of joining high temperature materials by melting them. But, polymer, or part of a powder is melted—the melting implies that this example of SLS does not fail to satisfy the condition to come under PBF.

In HSS, before PB is scanned by a thermal lamp, a radiation absorbing ink is deposited using an ink jet print head [8]. Use of the print head similar to that used in BJ3DP gives an impression that HSS comes under PB non-fusion, but the use of ink in HSS is not to bind powders but to facilitate subsequent melting of PB. HSS, therefore, comes under PBF.

In BJ3DP, binder is deposited on PB using an ink jet print head. The job of the binder is to hold powders together. If the binder is not capable to do that, a thermal lamp is used to improve the efficiency of binders but is not meant to melt powders [9]. SIS is same as BJ3DP in terms of the requirement of a jet print head, where inhibiter replaces the binder as a depositing material. The role of the inhibiter is to act as a negative binder.

If one process (SIS) uses negative binder and other (BJ3DP) uses (positive) binder, both processes can also be named as binder jetting process. But, this new name is not better than the previous name (PB non-fusion). Because, new name will not prevent another process (slurry bed) to be classified together with SIS and BJ3DP though slurry bed process does not use powder.

2.2   Role of Heat

Almost all PBP utilize heat as a necessity to realize the process. In BJ3DP, while binder jetting is used to shape a part, heat is utilized in post-processing to sinter the part. Similarly, in SIS, heat is used as a necessity in the form of post-processing sintering. In the absence of the post-processing [10], the part will not be formed.

In SLS, SLM and EBM, heat is used both for fabricating and post-processing. In SLM and EBM, controlled heat in the form of a point source is the only necessity to shape a part. Though for making useful parts, in-process and post-process heatings are used. In SLS, when the point heat source is used for shaping a part, post-processing heat treatment is the necessity to make a strong part.

3   Laser Powder Bed Fusion

LPBF, which is SLS and SLM, traces its origin back to 1986 when Carl Deckard, a master student at the University of Texas, filed a patent on this process, which was granted in 1989. The term 'laser' in SLS and SLM implies that a laser is used for processing, the term 'sintering' implies that powders are involved in the process. It infers that powder processing is done by the laser, which is used as a point heat source. 'Selective' implies that all powders are not processed by the laser simultaneously, i.e. powders are processed selectively when and where required. In conventional sintering, all powders are processed simultaneously. 'Melting' refers to a particular case of powder processing in which powders are completely melted. Thus, SLM refers to a case in which full melting occurs.

3.1   Description

In LPBF, the main aim is to make a layer of predefined geometry by fusing powders using an l-beam. The process follows the following sequence: (1) a substrate is lowered down to a depth equal to the layer thickness, (2) a powder layer is spread on the substrate, (3) the deposited powder layer is scanned by the l-beam to fuse powders at selected area. The sequence (1), (2), (3) is repeated until the fabrication is complete (Fig. 3).

In an initial stage of the process, powders are placed in a powder container and are protruded from the container by an adjoining piston. Adjacent to the container, a scraper is placed, which carries powders towards the substrate. The substrate is placed over a piston so its vertical position can be changed by adjusting the piston. Scanning mirror is used to scan the deposited layer on the substrate using an l-beam coming from a laser source.

In step 1 of the sequence, the piston of the powder container moves upward and the piston of the substrate container moves downward. The step gives requisite powders to be carried away by the scraper, and creates space on a substrate container for the powder to be deposited. In step 2 of the sequence, powders are deposited over the substrate and

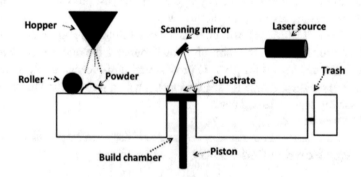

Fig. 3 Schematic diagram of laser powder bed fusion [11]

the position of the scraper changes to the right of the substrate. In the last step (step 3), deposited powders are scanned by an l-beam.

Above description shows the necessities of the process: (1) formation of PB, (2) consolidation of powders by an l-beam, and (3) a mechanism to repeat above listed first and second points.

Instead of using a substrate over a build platform, PB made directly over the platform can also act as a substrate. In this case, the final part is not needed to be cut off from the substrate as is not attached to the substrate. PB can also be formed by using a counter-rotating roller instead of a scrapper/blade and the powder feeding can also be accomplished by using a hopper instead of a powder container. From the hopper, a predetermined amount of powders fall in front of the roller, which then carries it away for deposition. The excess powder carried away falls into a trash that is equipped at the other side of the build chamber.

3.2   Parameters

Figure 4 shows schematic diagrams of powder layer processing with an l-beam. Figure 4a is for layer wise processing while Fig. 4b is for scan spacing and beam overlap.

3.2.1   Scan Spacing

Scan spacing is a separation between two consecutive l-beams. It is also called hatch spacing, which is the distance from the centre of one beam to that of the next beam (Fig. 4b). Scan spacing is directly proportional to the fabrication speed. If it is high, it will take less time for a laser to scan a layer. If it is low, a number of scans need to be performed to process the layer. Smaller scan spacing is required for making thin features.

To have a large scan spacing, a large laser spot size is required. Otherwise, there remains gap between two consecutive scans resulting in porosity. For processing with a larger spot size, higher laser power is required to supply necessary laser energy, which limits the maximum scan spacing obtained. To avoid any porosity formation at the boundaries of scans, some overlap (Fig. 4b), is made. Overlap is necessary because in a Gaussian beam, laser power at the centre of the scan is higher than at the boundary of the scan, resulting into melting at the centre while only heating at the boundary. Creating an overlap compensates this less heat irradiation at the boundary.

3.2.2   Scan Speed

Scan speed is the rate at which an l-beam scans a line on PB. Scan speed and laser energy density are related as follows:

$$E_v = \frac{P}{S_d \times V_s} \tag{1a}$$

Fig. 4 Diagrams for **a** layer
wise processing by an l-beam,
b scan spacing and beam
overlap [11]

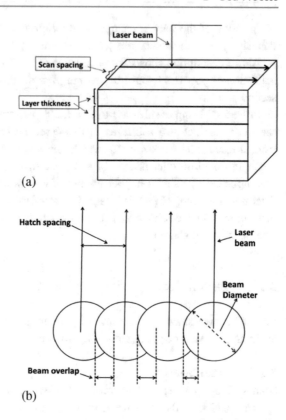

(a)

(b)

where E_v = *laser energy density, P = laser power, V_s = scan speed, S_d = spot size.*

  Equation 1a shows that at a high speed, laser energy density is low and may not be
sufficient to process PB. This can be compensated by increasing the laser power. However,
at a very high speed and laser power, the time is not sufficient for heat to diffuse across
the whole PB, it can lead to an insufficient melting and ablation of the powder. Therefore,
the equation holds good only within a limit that is determined by the types of materials
and other process conditions such as temperature and pressure of the environment. The
value of the scan speed used is in the range of 0.1 mms^{-1} and 15 ms^{-1}.

3.2.3 Layer Thickness

Layer thickness is the thickness of a powder layer used. Higher laser energy is required
for processing thicker layers. However, there is a limit to which laser energy can be
increased because supply of the high energy causes distortions on the surface and gives
rise to inaccuracies. This can be avoided by scanning the same surface twice using lower
energies.

  Laser energy density is given by the relation:

$$E_v = \frac{P}{S_s \times V_s \times L_t} \tag{1b}$$

where E_v = laser energy density, P = laser power, V_s = scan speed, L_t = layer thickness, S_s = scan spacing.

Equation 1b gives the laser energy density across the thickness of PB while Eq. 1a gave the energy density only on the surface of PB. Equation 1b holds when laser spot size is bigger than scan spacing. If the laser spot size is smaller than the scan spacing, Eq. 1b will consist of term spot size instead of the scan spacing.

Thin layer needs low energy density but furnishes dense parts with low surface roughness. For making a thin layer, small powder size is required. Using thin layers increases fabrication time. Low layer thickness means low shrinkage after melting by moving I-beam, which increases the dimensional accuracy and surface smoothness.

Selection of layer thickness depends upon the geometry of a part. When a curved object is fabricated layer wise, the layer being rectangular does not coincide with the contour of the curved object, it leads to gap on the side of the object, which is called staircase effect (Chap. 1).

The size of the gap depends upon layer thickness. For thinner layers, the gap is smaller but remains always present in a layer wise built. Therefore, the gap is minimized so the resulting contour is acceptable. But, decreasing the layer thickness has a limitation, which limits the minimization of the gap. To maximize fabrication speed without losing precision due to the staircase effect, the thickness of layers in a given built is optimized. For a vertical edge section of the part, higher thickness is selected while for a slope edge section, smaller thickness is selected.

The effect of both spot size (beam diameter) and hatch spacing together is given by Eq. 1c. Equation 1b is a special case of Eq. 1c when spot size equals scan spacing.

$$E_v = \frac{P}{V_s \times S_d \times L_t} \times \left[1 + \frac{(S_d - S_s)}{S_d}\right]$$

$$E_v = \frac{P}{V_s \times L_t} \times \left[\frac{(2S_d - S_s)}{S_d^2}\right] \tag{1c}$$

where E_v = laser energy density, P = laser power, V_s = scan speed, S_d = spot size, L_t = layer thickness, S_s = scan spacing.

3.3 Repair

Repair is a cost-effective measure taken to save: materials, energy for fabrication [12] and recycling, time and labor for original manufacturing. It is used in deposition process [13]: LENS [14], CSAM [15, 16], WAAM [17], AFSD [18], etc. But, it is not generally practised in bed process.

When a part is made from a customized material and the damage is not severe, repair can be employed in LPBF if following two conditions are met:

1. Damage should not be an obstacle for the damaged surface to be kept parallel to the platform or build plane, which will ensure no collision with a roller during powder spreading.

   For example, if all features marked as 1, 2, 3, 4, 5 (Fig. 5a) get snapped from a rectangular block ABCD, it is possible to keep the damaged surface AB parallel to the build plane and then repair the part. The front view of the block ABCD and surface AB is shown by rectangle ABCD and line AB, respectively.

   If these features are on an oblique surface (AB) of a triangular part (Fig. 5c), it is not possible to repair when any of the features on AB get damaged, as AB cannot be kept parallel to the build platform. If its orientation is changed to align it parallel, and fixtures are used to fix it parallel, it can compromise the accuracy of the repair.

2. Shape of a part should be such that when the part is fixed into a platform, the damaged surface must remain parallel to the platform. For example, lower surface CD (Fig. 5a) of the block ABCD is plane without having any complex feature on it, which is suitable to be fixed on the platform keeping the surface AB parallel to it (Fig. 6a). If the lower surface CD is tilted (Fig. 7a) or has a long thin feature (Fig. 7b), the part may not be fitted well to enable working of PBF system.

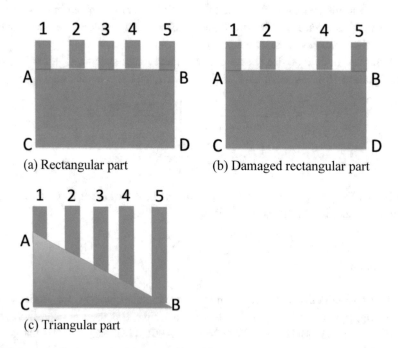

(a) Rectangular part   (b) Damaged rectangular part

(c) Triangular part

Fig. 5 Schematic diagrams of original and damaged parts

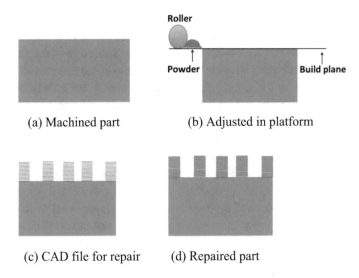

(a) Machined part   (b) Adjusted in platform

(c) CAD file for repair   (d) Repaired part

Fig. 6   Four stages of a repair

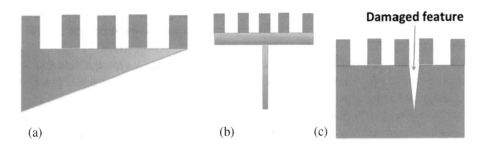

(a)   (b)   (c)

Fig. 7   Non-repairable parts **a** oblique lower surface, **b** thin feature in lower surface, **c** deep damaged feature

Figure 5a shows an original part having five features while Fig. 5b shows the same part having one feature numbered 3 (feature-3) missing. Features are blocks of equal height having the same cross-section. To repair the damaged part (Fig. 5b), the missing feature needs to be regrown at the same location of the part.

It is not possible to regrow a feature selectively unless in an exceptional case when the height of the feature is approximately equal to the layer thickness—when the height of feature-3 is equal to the layer thickness, the part can be repaired selectively.

For repair, what is required is the powder in a vacant space between feature-2 and feature-4, and selective consolidation in the vacant space as per the contour of feature-3. The powder is brought by spreading it using a roller, which fills up the vacant space. As the upper surface of features-1,2,4,5 will come in contact with the roller, it is assumed that these features are strong enough not to break due to friction with the moving roller. This again brings limitation to the type of damage a part can have to make it repairable.

As scanning is optimized for one layer thickness, the formation of the feature will not face the problem of inadequate bonding between the base of the feature and a rectangular block. If the height of the feature is bigger, scanning will not furnish adequate bonding.

This example of a missing feature having height equal to a layer thickness is unlikely to occur but it, nevertheless, shows the problems while repairing. The problem is not the presence of a missing feature (feature-3), but the presence of other features (features-1,2,4,5). The other features block the access of a coating device (roller) to the site of the missing feature. In this case, these features have not blocked the access because all these features are of equal height, which does not prevent the roller from rolling. Besides, it is assumed that there is a need of deposition of only one layer as all features are of one layer height—it means the roller has a role to play. If the features were of more one than one layer height, the rolling had no utility as the consolidation after rolling could not have been achieved—the consolidation needs to happen layerwise.

If features are required to be made from several layers, all features must be removed by machining to build all new features instead of just one new feature. Figure 6a shows a machined part without any features, which can be fitted well into the platform (Fig. 6b) because surface AB is parallel to the build plane. Lower parallel surface CD will help it fixed. A CAD file having all features sliced (Fig. 6c) is used to build features to make a final part (Fig. 6d).

Need to remove four features for repairing just one feature brings a question mark on the efficiency of PBF for repair, but economically and resource-wise it is better than both: not using the damaged part, and opting for a new part.

If an original part is damaged severely (Fig. 7c), where there is a triangular cut reached inside the part, repair may not be an option. If it is machined parallel to AB as done earlier (Fig. 6a), approximately 3/4th of the part is machined out. If it is machined parallel to BC so the repair can be done by changing the orientation, approximately 2/5th of the part is removed.

If it is machined from side AC, approximately 3/5th of the part goes. There is no option to machine from side CD as approximately 4/5th of the part is removed as well as all five features are separated. In all cases, remaining parts after machining are not large enough to bring significant difference between repair and new fabrication, questioning the reason for repair [19].

Since a damaged part needs to meet some conditions before it can be repaired, it could be better if the design of a part could be changed keeping in view of the future need for repair [20].

3.4   Methods for Increasing Fabrication Rate

Increasing fabrication rate has a direct consequence on decreasing product cost and increasing production efficiency [21]. There are a number of methods by which a high rate can be achieved. These methods are:

3.4.1   Increasing Layer Thickness

Increasing layer thickness decreases the number of layers required to make a part [22]. Therefore, the number of times powder is deposited to make all layers decreases, which in turn decreases the fabrication time and energy consumption [23]. However, this method has a limitation—increasing it more than a certain thickness does not let the effect of melting reach the base of the layer [24]—powders either bind by solid state sintering if build continues for long or remain unaffected.

Thus, there is a limitation to which the rate can be enhanced by this method [25]. This method is applicable to all layer based processes [26].

3.4.2   Dividing Processing Area

Dividing a processing area into a number of zones and processing all zones simultaneously increases the rate [27]. The layer is processed by a number of beams parallelly (Fig. 8a). Fabrication time is proportional to the total area divided by the number of beams when each beam scans an equal area. Therefore, the time decreases with an increase in the number of beams—this method can lead to mass production [28].

The reason for an increase in the rate is the addition of efforts. The effort is to join materials, which is limited by size of a machine as a small machine can accommodate only a small number of beams. Besides, this method is limited by managing a sequence of operations involving many beams. It is further limited by sustainability—if one beam fails, the beam needs to be fixed before all beams can be used together.

The method is again limited by cost effectiveness and logic—why not making many machines each equipped with a single laser than a single machine having many lasers. As the percentage of the cost of the laser is high to the total cost of a simple laser based machine, it would not be much expensive to have many machines than a single big multi-laser machine. Presence of many machines will allow to work with different materials in separate machines and make different products.

The method has limitation for making a small part where the size of the boundary between two zones is not significantly smaller than the area of the zones when two zones are processed by separate beams. Different sizes of melt pool will be formed by different beams that have small variation or noise. For a small part unlike a big part, this variation will be significant.

For making a number of small parts in PBF that is equipped with a number of beams to employ parallel scanning (Fig. 8b), the task of processing will be shared equally among each beam [29]. It will increase the overall fabrication rate but will neither increase the

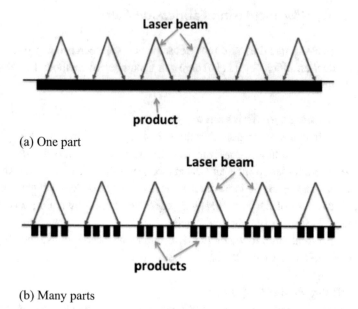

(a) One part

(b) Many parts

Fig. 8 Application of a number of beams for making **a** one part, **b** many parts

rate for a single part nor require synchronization of beams to process the part (as a part is processed by only one beam). The single part thus fabricated in a multi-beam setting will not be different from a part if it is fabricated in a similar setting but using a single beam.

Figure 8a shows beams from six laser sources contribute to make one part while Fig. 8b shows beams from six laser sources, they each scan equal space (1/6th of space) and make four parts each.

For scanning a layer by parallel beams, if their heat affected zones are overlapping, the development of thermal stress across the layer will be minimal. Thus, parallel scanning increases the production rate while minimising the thermal stress.

3.4.3   Optimizing Process

If high strength is not required for all sections of a part, the section that does not require the high strength will be scanned with a fast scan speed, which will increase fabrication rate [30]. Alternatively, the section can also be scanned by less overlap between two adjacent scans, which will decrease the number of scans required to cover the section, causing a decrease in scanning time. Besides, if instead of high strength an optimum strength is required, there can be several build rates [31].

Hence, optimizing process parameters as per the requirement of different sections will not only pave the way for increasing the rate but also decrease the energy required. The type of gas used in a processing chamber can be one of the process parameters, as some gas allows to obtain high density at high rate, and thus help increase the fabrication rate [32].

If various beam diameters are available, using the large diameter will decrease the time to scan the whole layer and increase the fabrication rate [33].

3.4.4   Changing Orientation

A part can be made in many ways by changing its orientation (changing the angle between its features and build direction) [34]. If changing the orientation results in a decrease in number of layers required for the completion of the built, fabrication time will decrease. However, this method has limitations: a decrease in mechanical properties (heat accumulated on a top layer will not be same in all orientations), need for support structures [35], change in the grain size (length of a columnar grain will not be the same in all orientations), property changes [36], and a decrease in surface roughness and geometrical accuracy.

The effect of the orientation change is shown in Fig. 9 for making a hollow rectangular part, which is open at one side surface (to let powder drain out after fabrication). There will be an increase in the fabrication rate with a decrease in the height of the build, but it will result in a big overhang (Fig. 9a). Changing the orientation by 90° (Fig. 9b) will increase the height of the build, which will decrease the rate. But it will decrease the size of an overhang required to be built and, therefore, be free from difficulties for making a big overhang. Hence, the orientation (Fig. 9a) furnishes lower surface finish for a big area (underneath the big overhang) while the orientation (Fig. 9b) furnishes lower surface finish for a small area (underneath the small overhang). Besides, making big overhang without support structure means losing the planarity of the overhang surface, causing a decrease in the dimensional accuracy.

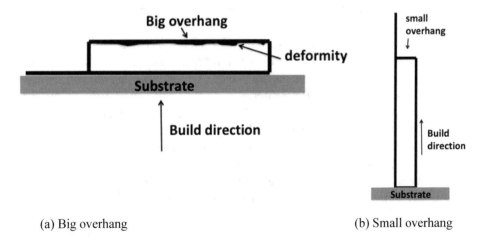

(a) Big overhang   (b) Small overhang

Fig. 9 Different orientations: **a** big overhang, **b** small overhang

3.4.5 Increasing Scan Speed

A scan speed can be termed low or high depending upon materials and experiments. A low scan speed of 689 mm/s is considered high when it gives porous parts [37].

Increasing scan speed increases the fabrication rate. Though, the highest speed used is limited by the following:

1. Evaporation of powders rather than melting—at high speed of a high power l-beam, top section of a powder accumulates high amount of heat, leading to its evaporation.
2. Formation of a longer melt pool that breaks into shorter pools, giving rise to discontinuity, which is due to Rayleigh instability that states if the length by width ratio is more than π (a critical value depending upon interface), the pool will break. Longer pool will no longer be long if it solidifies on the way, i.e. if solidification rate is faster than or equal to the scan speed. For an aluminum alloy scanned at 1 m/s, the solidification rate is almost the same [38]. Cooling rate remains high in an experimental setup giving high solidification rate—this is one of the reasons why Rayleigh instability is not the main problem for consolidation.

   Which one of the above two reasons can be the main reason for failure at high scan speeds (>100 m/s)? At a certain high speed, only evaporation takes place and there will be no melt pool. In the absence of any melt pool, the first reason is the cause of failure. Increasing pressure of a processing chamber will let increase the speed as the pressure suppresses the evaporation—more pressure means higher speed. There is a limit beyond which application of the pressure does not work—either due to its effect on the melt pool, such as depression, or due to the safety aspect involved with working at high pressure.

   Increasing the speed further without increasing the pressure will bring forth a result no more different than evaporation unless laser power is not increased proportionally. In the absence of increased laser power, there will be neither evaporation nor melting but just heating. With a decrease in speed, there will be a certain speed at which evaporation will stop and melting will start. If solidification rate is similar to the speed, this mechanism will be able to furnish a part. If the solidification rate is low, two cases will happen: (a) part formation will fail due to Rayleigh instability, (b) part formation will not fail if some measures are taken.

   One measure is increasing cooling rate by decreasing the temperature of a substrate, which will increase solidification rate but a part may still fail due to the formation of cracks. Another measure is to employ laser pulse (~ms in order to prevent evaporation) instead of continuous l-beam, which will create smaller pools. Applying the mechanism at high speed will leave few options to exercise for making a defect-free part, decreasing the freedom to vary microstructures and properties.
3. High speed requires high laser power to maintain required laser energy density for melting [39]. High laser power means high cost of a machine. Working with high

laser power means to manage high heat generated, which requires upgrading a device (laser, mirror, etc.) and incorporating new cooling device to safeguard the machine.

In EPBF, there will not be any damage to scanning devices at high speed, but the lifetime of electrodes is diminished at the constant generation of high power.

3.4.6   Adopting Areawise Scanning

Pointwise scanning means scanning using a point source (such as laser spot size), which takes time for processing a layer—point source has to process one point on PB then it has to move to another point. In fast pointwise scanning, the processing moves fast but has to cover many points, which is the limitation.

Changing from pointwise to linewise [8, 40], or areawise scanning is another method to increase the rate. In the linewise scanning, a number of l-beam sources (diode emitters) are fixed together on a horizontal line, and the line moves to process layers. In the areawise scanning, all heat point sources are fixed in an area over PB and process the bed simultaneously without either movement of heat source or deflection of heat radiation [4].

4   Electron Beam Powder Bed Fusion

A schematic diagram of EPBF is given in Fig. 10, which shows following requirements to realize the process: (1) beam generation, (2) beam manipulation, (3) vacuum chamber and (4) powder bed [41], which are described further.

Fig. 10 Schematic diagram of electron beam powder bed fusion

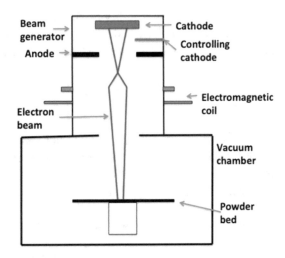

4.1   Beam Generation

E-beam is required to melt and sinter powders: melting for shaping while sintering for fixing the powder at PB when required [42]. The beam is generated by electrically heating a cathode made from a material having high melting point and low work function. Due to the high melting point, the cathode does not melt and degrades during electron generation while due to the low work function, electrons detach from the cathode at a low applied voltage. Work function is a physics term that states how much voltage needs to be applied to a material to detach an electron from its surface. Lower the work function, better the material for e-beam generation.

Tungsten (W) and Lanthanum Hexaboride (LaB$_6$) are used as a cathode material, each of which has melting point and work function as 3422 °C, 4.5 eV and 2210 °C, 2.5 eV, respectively. W is used because of its high melting point while LaB$_6$ due to its lower work function and higher lifetime as a cathode [43]. Besides generating beam by heating the cathode, which is called thermoionic emission, it is generated by creating plasma between the cathode and the anode [44]. The beam, thus produced, is used in AM [45].

Generated electron (charge $= -1.6 \times 10^{-19}$ C, mass $= 9.1 \times 10^{-31}$ kg) moves towards anode. The movement of electrons constitutes a beam. Higher the number of electrons, higher the current of the beam. Thus to have a high current, a large number of electrons need to be generated. While to have a low current, either a smaller number of electrons need to be generated or some electrons need to be stopped from reaching the anode. For stopping the electrons, placing another cathode just before the anode serves the purpose as the cathode displaces some of the electrons before they reach the anode, which decreases the current. Anode is made perforated so the remaining electrons do not stop at the anode but passes through it so that they can be manipulated for processing PB.

The speed of an electron depends on the voltage applied between the cathode and anode. Higher the voltage, higher the velocity of the electron as per Eq. 2a:

$$v = c \times \sqrt{1 - \frac{1}{\left(1 + \frac{e \times V}{m \times c^2}\right)^2}} \tag{2a}$$

where v, e, m are the velocity, charge, and mass of an electron, respectively. V is an applied voltage and c is the speed of light [46]. For V of the order of kV, the speed of the electron is of the order of c, which brings relativistic effect in m, which can not be neglected if voltage is more than 100 kV [47]. Thus, m will increase by $\frac{1}{\sqrt{1 - \frac{v^2}{c^2}}}$.

The power of an e-beam having current I is given by Eq. 2b:

$$P = V \times I \tag{2b}$$

Thus the power can be increased either by increasing the voltage or current. With an increase in the voltage, electrons move fast and go deep inside PB while with an increase in the current, more electrons reach PB. Consequently, for the same value of power, there will be a number of different physical effects on PB depending upon the value of V and I. This is the reason why V and I (rather than P) are independent variables in EPBF.

To increase P, increasing V rather than I is a better option. Since, with an increase in I, the number of electrons increases giving rise to an increased charging of PB causing a disturbance on PB. Increasing V causes an increase in the momentum of the electron but due to a huge difference in the mass of the electron and of the powder, and small beam–powder interaction time, the impinging electron is not able to displace the powder, causing no disturbance on PB. Though the high velocity of the electron causes an increase in the mass of the electron due to relativity, the increase is still miniscule to move the powder position.

4.2   Beam Manipulation

An e-beam generated from a cathode and passing through a perforated anode will strike a limited area of PB, i.e. an area equivalent to the beam spot size, which is of the order of millimeter or less. It will not help make parts bigger than the spot size unless e-beam generator moves relative to PB, or vice versa. Making parts by these relative movements can be helpful for investigating the effect of beam parameters on materials. But this has limited utility in a commercial application for the lack of accuracy and speed.

Another option is to displace e-beam from its original path so it will cover wider area of PB and be able to make bigger parts. Since an e-beam consists of charged particles, it can be displaced by an applied electric field. Besides, the charged particle moves, which creates a magnetic field that can be displaced by an applied magnetic field. Hence, an e-beam can be displaced both by an applied electric and magnetic field, which is not possible for a laser beam as it does not consist of charged particles. These show an electromagnetic field that has both electric and magnetic field components can be suitable for e-beam manipulation.

An electron of charge e moving with a velocity v will feel a force F, called Lorentz Force, in an electromagnetic field as per Eq. 2c:

$$F = e \times (E + v \times B) \tag{2c}$$

where E and B denote an electric field and a magnetic field, respectively. F acts normal to the plane made by v and B. Figure 11a shows x-, y- and z-directions in cartesian coordinates.

To comprehend the role of the magnetic field on a moving electron, the electric field is considered negligible, which changes Eqs. 2c–2d.

$$F = e \times (v \times B) \tag{2d}$$

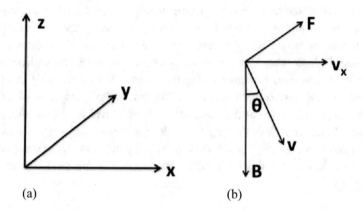

Fig. 11 **a** Cartesian coordinates, **b** Lorentz force F on an electron moving with velocity v in a magnetic field B

Expanding Eq. 2d in terms of its components in x-, y- and z-directions, it can be shown by Eqs. 2e, 2f, 2g.

$$F_x = e(v_y B_z - v_z B_y) \tag{2e}$$

$$F_y = e(v_z B_x - v_x B_z) \tag{2f}$$

$$F_z = e(v_x B_y - v_y B_x) \tag{2g}$$

To find a resulting force on an electron moving with a velocity v in the z–x plane and along a direction at an angle θ with B (which is in z-direction) (Fig. 11b), following information is used in Eqs. 2e, 2f, 2g:

$B_y = 0$, $B_x = 0$ since B is in z-direction, $v_y = 0$ since v is along the z–x plane, it is found $F_y = -ev_x B_z$ while $F_x = 0$, $F_z = 0$.

Since, $v_x = v \sin \theta$, $B = -B_z$

$$F = F_y = -ev_x B_z = ev B \sin \theta \tag{2h}$$

Equation 2h states the direction and magnitude of Lorentz force acting on a moving electron. The force is in y-direction and acting on the z–x plane (Fig. 11b).

The consequence of this force on a moving electron is known by observing the direction of the force. Since the magnetic field acts in a circular direction on a moving charge particle (electron), it acts in a clockwise direction on the electron to cause resultant force in y-direction. It is similar to the rotation of a screw that if rotates in a clockwise direction from v to B through a smaller angle in the z–x plane (Fig. 11b) causes an advancement in the positive y-direction. Vice versa, if there is a force in y-direction, it means there was a rotation of electron clockwise. Observing the direction of F in Fig. 11b, it is known whether electron direction is clockwise or counter-clockwise. Clockwise implies displacement of the electron towards the magnetic field while counter-clockwise implies away from the magnetic field.

In the present case, as per Fig. 11b, displacement of electrons is towards magnetic field. How far an electron will displace depends upon the magnitude of v and B. If B is large or there is a strong magnetic field, force will be large enough on the electron to align it with the magnetic field. With an alignment, angle θ will be zero and, as per Eq. 2h, force will be zero, causing the electron to move indefinitely parallel to the magnetic field. This concept is utilized in focusing the beam in EPBF.

Using $\theta = 0$ in Eq. 2h, $F = 0$, it gives another inference that when magnetic field is applied parallel to the beam, it has no effect on the electron while for any other value of θ, the beam is deflected. It implies that the magnetic field can not be used to accelerate an electron along optical axis, acceleration can only be accomplished by an applied voltage.

In case, magnetic field B is reversed so $B = B_z$, it will change the equation to $F_y = -ev_x B$, which is acting in the negative y-direction normal to the z–x plane implying a counterclockwise motion of vector v. It will make the angle θ bigger.

This shows that by reversing the magnetic field direction, electrons will be displaced away from the magnetic field. This will help the beam to move away from its original path and scan wider length. By fast fluctuating magnetic field direction from positive to negative, the beam direction will change fast while by changing the strength of the magnetic field, the beam velocity on PB (scan speed, Fig. 12) can be changed. This shows that there are two ways by which scan speed on PB can be changed—by changing either applied voltage or magnetic field. For scanning along more axes, more electromagnets (as sources of magnetic fields) are required. This concept is utilized in scanning the beam.

Figure 12 shows scanning of PB along a line using an e-beam moving with a velocity v. Scan speed s does not depend upon the beam velocity but depends upon the speed with which the beam is deflected towards optical axis, which depends upon the magnitude of magnetic field. If a high magnetic field is applied, the beam travels fast on PB imparting

Fig. 12 An e-beam with velocity v scans powder bed with scan speed s

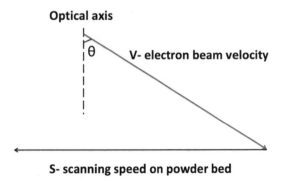

less number of electrons on the bed. If the field is low, scanning speed will be slow imparting a high number of electrons on the bed, resulting in a high effective current on the bed, implying higher energy density applied.

For a constant applied voltage (60 kV in Arcam EBM systems), changing applied current is a means to control the number of electrons on PB. If scan speed is set higher, resulting in a lower number of electrons, the applied current can be increased to compensate a decrease in electron numbers, resulting in no change in the current density with an increase in scan speed. Thus, the scan speed and applied current can be adjusted to maintain desired energy balance on the bed. Finding the right energy density by making an adjustment towards higher scan speed will result in a high production rate.

4.2.1 Focusing of Beam

Focusing of an e-beam is done by applying magnetic field parallel to an optical axis along which the beam is expected to travel. The direction of the magnetic field, optical axis and the velocity of the electron is shown in Fig. 13a. E-beam after being accelerated by an applied voltage does not exactly move in a straight line along the optical axis but moves in a curved path due to the presence of other electromagnetic fields (Fig. 13a). If an e-beam entering the magnetic field makes an angle with the optical axis, it will be deflected by the field (Fig. 11b) and will become parallel with the optical axis (Fig. 13b).

Applying a magnetic field parallel to an optical axis helps achieve focusing of a beam but it can also be achieved by applying the field perpendicular to the axis (Fig. 14). If the beam is deflected from the optical axis, it can be brought back to the same path by applying B perpendicular to it (Fig. 14a), in which case, the beam is making an angle

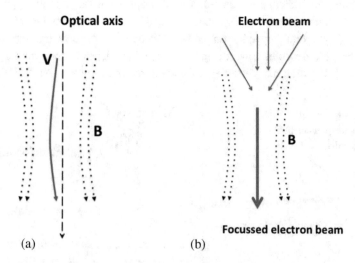

(a)   (b)

Fig. 13 An electron moving with velocity v in a magnetic field B **a** approximate beam path [48], **b** focusing of e-beams entering the field at different angles

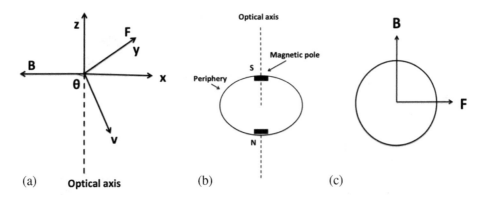

Fig. 14 Focusing of e-beam by deflection **a** directions of B, F and v in cartesian coordinates, **b** position of electromagnet on the periphery of e-beam column, **c** directions of B and F

θ with B. The magnitude of the field should be set to a particular value so the resulting force will deflect it by an angle $\theta - 90°$ and align it with the axis, causing focusing of the beam.

In order to execute it, two magnets are attached on the periphery of an e-beam column—north pole of one magnet while south pole of another magnet is facing the periphery (Fig. 14b), which generate magnetic fields. The direction of the field and of the force are perpendicular to each other (Fig. 14c). These magnetic poles displace the beam along x-direction. In order to displace it along y-direction, two more magnetic poles need to be set on the periphery normal to the fitted poles. These four magnetic poles will force align an e-beam making any angle with the optical axis, causing focusing of the beam.

4.2.2 Beam Shape Control

Beam during scanning may not retain its original shape and deforms to various shapes. Figure 15a shows a beam of original spherical shape that deforms resulting in elongation along AB while compression along CD (Fig. 15b). The beam can be brought to its original shape if compressive force acts along AB, and pulling force acts along CD (Fig. 15c). Electrons at point D of the beam need to be deflected outwards while electrons at point A need to be deflected inwards to bring them back to their original position. As per Fig. 14a, the former deflection can be achieved by applying B along x-axis, orthogonal to the optical axis (z-axis) while the latter inward deflection can be achieved by applying another magnetic field in the negative direction along y-axis—this requires an arrangement of four magnetic poles.

This can be achieved by a quadrupole magnet (Fig. 15d), which applies the force exactly in the same direction as required [48]. The figure shows four poles are placed alternatively, magnetic lines are travelling from N to S creating two opposite forces: one

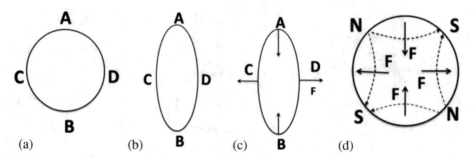

Fig. 15 **a** Original beam shape, **b** deformed beam shape, **c** direction in which force needs to be applied for regaining the shape, **d** quadrupole magnet as a device to apply required force

in x-axis and the other in y-axis. For a higher degree of deformation, a higher number of poles of equal number of similar poles are required.

4.3   Vacuum Chamber

A vacuum chamber is an essential part, in which e-beams travel and interact with PB. In the absence of vacuum, electrons will interact with air molecules and ionize them, thus lose energy, speed and direction during travel, causing difficulty in processing PB. In Arcam systems, a vacuum of 10^{-5} mbar is maintained, which is sufficient to allow unhindered movement of e-beams. Besides, the vacuum enables reactive metals to be processed without getting reacted by oxygen or nitrogen gas [49].

Vacuum has disadvantages—it decreases melting point of metals and increases its evaporation rate. For an alloy made up of two different metals of different melting points, the evaporation rate of lower melting point metal will be higher resulting in its higher loss, causing a change in alloy composition. This is noticed in Ti6Al4V alloy where the higher loss of Al changes the composition.

Interaction of e-beam with powder leads to the transfer of electrons to the powder causing its charging. Accumulation of charges leads to coulombic repulsion between same charges, which destabilizes powders' positions and bring disruption in a layer formation. When charges are high, the resulting repulsion is higher than the inertia of powders leading to their displacement. Presence of helium gas (10^{-3} mbar partial pressure in Arcam systems) in the vacuum chamber mitigates this effect. Helium has low atomic number, and at a pressure of 10^{-3} mbar, number of helium atoms is not large enough to disturb the beam. But these numbers are big enough to take substantial amount of charges away from charged powders by being present in the vicinity of these powders, causing mitigation of disruption of PB.

4.4   **Powder Bed Processing**

After layer formation, if powders get attached to a substrate or previously deposited layer, the incoming electrons will not be able to accumulate on the powder to disturb it. Attaching the powder can be done either by sintering or melting it. Melting or partial melting of powders cannot be an option as all powders that are not melted by e-beams need to be detached and removed from the final part. As melting is the way for fabricating, the powders that are melted contribute to fabrication. Therefore, the powders are not supposed to be melted for any purpose other than contributing to the fabrication.

This leaves another option i.e. sintering to be exercised for making connection among powders and with the substrate [50]. Sintering implies that the powder will join another powder or substrate by forming a neck through atomic diffusion. Formation of neck increases electrical conductivity of the layer and enhances contacts between the substrate and powders, which helps dissipation of charges to the ground, resulting in removal of cause (charges) responsible for dislodging of powders. A minimal size of neck obtained after some moment of sintering is sufficient for charge transfer and stabilization of PB. Since the role of the neck is to act as a conduit to transfer charge rather than to act as a fixture to hold a powder, a large powder [51] because of its higher inertia rather than a smaller powder has a higher chance not to be displaced during charging. This makes the larger powder more suitable for EPBF.

Sintering is accomplished by the application of heat with the help of heaters and e-beams. A heater attached under a substrate is used to increase its temperature. When a layer comes in contact with it, powders of the layer are sintered. The low temperature of the heater increases the sintering time, which decreases the fabrication rate. While the high temperature increases the sintering of un-melted powders of previously deposited layers, resulting in difficulty in removal of the powders from the part, which compromises fine features. This can be avoided by the use of a low heater temperature combined with heating the layer with the help of an e-beam. This will ensure a no decrease in the fabrication rate and the non-occurrence of over-sintering. For increasing the local temperature of the layer, high power e-beams with high speed are scanned over whole layer several times, which causes necessary sintering. An e-beam moving at the high speed does not impart enough electrons at a point to cause disturbance on the layer.

E-beam for the purpose of sintering is used at several stages—on the substrate (as a substitute for the heater), on the deposited and processed layer (for helping the heater), and after processing all layers for annealing. Any of these stages increases the temperature of a processing chamber. A hot chamber gives advantages: decrease in thermal stress, cracks, porosity, and disadvantages: decrease in strength in several alloys, grain size increase, etc.

5   EPBF and LPBF

EPBF and LPBF are compared as follows:

5.1   Powder

In EPBF, powder needs to be electrically conductive to facilitate dissipation of charges from it to surroundings. This limits the types of powders that are suitable to be processed, therefore, mostly metallic materials are used. Powders that form oxide or nitride layers on their surfaces due to contamination with gases diminishes their suitability as these layers have lower electrical conductivity.

In EPBF, powder can be displaced due to charging, while LPBF does not have such limitation. This implies big powder is suitable in EPBF while smaller powder can be used in LPBF [52]—this leads to higher surface roughness in EPBF [53].

In LPBF, there does not exist a yardstick (such as electrical conductivity existing in EPBF), the lack of which makes the materials un-processable.

5.2   Beam

In the case of conversion of electrical energy into beam (electron or laser), known as wall-plug efficiency, the rate of conversion for e-beam is about 95% while for l-beams used in LPBF, it is from 5 to 20% for CO_2 laser, up to 20% for Nd: YAG laser, from 30 to 40% for fibre laser, up to 70% for diode laser [54, 55], showing the higher efficiency of e-beam over l-beams.

5.3   Beam–Powder Interaction

E-beams are made up of electrons that when irradiated on metal powders get deflected by free electrons surrounding metal atoms. If the power of the e-beam is high, the e-beam will not be deflected by free electrons but move deeper inside the metal until decelerated by a lattice of atoms and gets stopped. The interaction causes the lattice to vibrate and generate heat. Thus, the kinetic energy of moving electrons is transferred into heat energy, responsible for melting. If the size of the atom is big (higher atomic number), there is a less chance for the e-beam to escape. Thus, a bigger atom size lets higher percentage of kinetic energy be transferred, which increases the efficiency of the process. For an applied voltage of the order of kV, the size of e-beam impingement in a powder is of the order of microns.

L-beam is made up of photons that are neutral. When it irradiates metal powders, photons interact with free electrons surrounding atoms. For a high number of free electrons, a photon has a chance to be reflected. For an l-beam of high intensity irradiating a metal having fewer number of free electrons, incoming photons interact with bound electrons, which causes the electrons to re-radiate or interact with lattice atoms, causing vibration. L-beam is assumed to be high intensity because if it has low intensity, it will less likely reach to the bound electrons. The metal is assumed to have a small number of free electrons, because if it has high number of free electrons, the beam will be more likely to be deflected by these electrons without reaching the actual metal surface comprising of bound electrons.

Photons deflected by bound electrons have less chance to escape (than the photons deflected by free electrons) and will reflect internally until it loses energy by interacting with the lattice. Thus incoming laser energy is transferred to metal powders as heat energy. When incoming photon reaches bound electrons without being deflected by free electrons, it is able to transfer its energy. This distance inside the powder (penetration depth) is about few nm (Fig. 16).

This shows that l-beam–material interaction is a surface phenomenon while e-beam–material interaction (penetration depth—few μm) is a relative bulk phenomenon.

While an l-beam is prone to be reflected from the metal powder, there is no such occurrence with e-beam, making the e-beam more efficient in terms of transfer of energy to metals (as well as in terms of generating it from the electrical energy). The comparative inefficiency of l-beam is somewhat compensated in PB processing when the reflected beam has a chance to be directed towards another powder and gets absorbed. This depends upon the surface roughness of powders, gap between powders and the angle of incidence.

Metals such as copper, aluminium, silver that have high reflectivity are not suitable to be processed in LPBF but in EPBF. These reflective metals on developing an oxide layer

Fig. 16 Incidence of l- and e-beams at metal surface

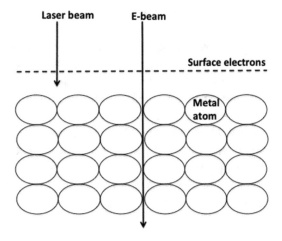

such as alumina layer on aluminium are no longer suitable for EPBF because of the charge accumulation, but they are suitable for LPBF because of their decreased reflectivity. This example demonstrates what is good in one process may not be good in other, and vice versa, implying both processes can complement each other.

5.4   Parameter

L-beam moves with a constant speed, i.e. the speed of light while e-beam has varied speeds depending upon the applied voltage. Applied energy density in EPBF can be changed by changing the speed of the e-beam while there is no such option in LPBF as the speed of light is a constant.

Laser power is an independent quantity. For a given setting of laser power in LPBF, it does not change with a change in other experimental parameters such as scan speed and scan strategy. In EPBF, power may change with a change in scan speed or scan strategy. This is the reason why keeping power constant and changing other experimental parameters give useful information in LPBF while the same gives misleading information in EPBF, if all details are not taken into account.

LPBF can be carried out using an l-beam of different types of wavelengths such as YAG laser, fibre laser, etc. Different wavelengths have different absorptance for different materials. There is no such provision in EPBF that it can be carried out using an e-beam of a particular wavelength. Since an e-beam accelerated at an applied voltage has a particular wavelength (de Broglie wavelength) but during EPBF, the speed of the beam varies bringing variation in wavelength, it implies that there is no EPBF with a fixed wavelength though there is LPBF with a fixed wavelength.

5.5   Implication of Vacuum

Vacuum increases the evaporation rate of materials and changes the composition of a processed material, which implies that LPBF (without vacuum, with inert gas) will cause lesser composition change than EPBF does. Vacuum has advantage as there is no gas in a chamber and therefore no gas will be absorbed by liquid metal pool, absorption of the gas is a source of porosity.

LPBF has an edge over EPBF to process the material by increasing atmospheric pressure in the chamber. High pressure will decrease porosity, surface roughness [56] and sublimation. Besides in LPBF, gas flow inside the processing chamber can be managed to change surface roughness and quality [57]. Flow parallel to PB is used to remove particulates from the chamber and is used as a parameter to vary properties [58].

5.6   Scanning

EPBF works using electromagnetic coil based scanning system that enables scan speed of the order of thousand m/s, while LPBF works on mirror based scanning system that provides scan speed of the order of ten m/s. EPBF, due to its higher scan speed, furnishes distinct advantages over LPBF, e.g. potential for higher production rate, mitigation of thermal stress by use of simultaneous parallel beams, utilization of beam as a source of an additional heater, etc.

In EPBF, high scan speed combined with high beam power enables transfer of high energy density at a faster rate leading to fast melting of the whole layer, this fulfils one of the tasks for higher production rate. LPBF has lower scan speed, and does not match EPBF in this aspect. However, there is a limit by which scan speed in LPBF or EPBF can be increased. At higher scan speed, there will be small beam–powder interaction time (of the order of nano second for a beam size of 100 μm moving with 1000 m/s). This will cause ablation [59] i.e. non-melting of powders, generating evaporated materials interfering with vacuum and beam (particularly in EPBF), and pressing the molten material by recoil pressure.

Scan speed more than a certain speed can not be employed. As a result, a beam power more than a certain power can not be employed to scan. This is the reason why the laser power of several kW is available but there is no LPBF system equipped with that. Nevertheless, higher power can be utilized by distributing it among several beams with a restriction that the power in a beam should not exceed the power limit. If the processing chamber is hot, as in EPBF, still lower power (or applied energy density) is enough to raise the temperature of the layer.

6   Non-beam Powder Bed Fusion

Beam PBF makes complex parts but is expensive because beam is expensive. If the beam is replaced with an alternative heat source, the process will not be expensive. A resistance heater can do the job [3] but the heat from it is not as directed as a beam is. The beam has an advantage that it can be deflected and the whole powder bed (PB) can be raster scanned. The heater does not give such convenience; it needs to be moved relative to PB or vice versa. Alternatively, a number of small heaters need to be arranged all over close to PB so by switching them on and off, the whole PB can be selectively heated, which will lead to a process that is free from a beam [4].

Since the aim of applying heat is to join powders, spraying them with glue (or ink or binder) can fulfil the aim as well [60]. Thus, there is another process where both the beam and heater is done away with an ink jetter or a binder jetter.

Beam PBF is widely researched because a beam provides high energy density and resolution. The high energy density allows to process metals and ceramics, while the resolution allows to introduce details in a part. But, the process is slow and requires much energy. These demerits drive to search for other processes not based on high energy beams.

Non-beam heat sources are infrared lamps, concentrated microwave energy and resistance based heaters. Microwave energy controlled by localized microwave heating applicator is not considered a beam because it, unlike the beam, can not be deflected. Besides, the applicator needs to be placed near PB surface while e-beam or l-beam sources do not have such compulsion [6].

6.1   Heater Based Sintering

If a heater instead of a beam source is placed above PB, radiation from the heater will not converge to a point below the heater on PB. With an increase in separation between the heater and PB, the heat energy density at the point will further decrease. Therefore, if the aim is to melt powder, the heater must be in proximity to PB, which will damage the heater itself if not shielded. Though it can work well with a low melting point material such as polymer for which less heat-induced risk is involved.

A heater needs either to move to cover the whole area or be present everywhere so the whole area will be under its reach. If the heater needs to move, it reminds the movement of an ink jet printhead that covers the whole area without needing the quality of deflecting a beam. An ink jet printhead consists of a number of ink jets. Thus, if the printhead instead consists of a number of small heaters, it will serve the purpose. This is the concept behind selective heat sintering (SHS) (Fig. 17). If the heater needs to be present everywhere, a number of small heaters need to be fitted in a cover or mask covering the whole area. This reminds the presence of a number of small mirrors in digital light processing. Selecting some of the heaters switched on while others off will help print selectively. This is the concept behind micro heater array powder sintering (MAPS).

Fig. 17 Schematic diagram of selective heat sintering

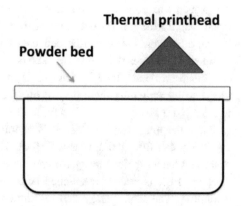

Thermal printhead

Powder bed

6.2   Localized Microwave Heating Based AM

Microwave is an electromagnetic radiation of wavelength from 1 mm to 1 m while laser used in AM is of the order of μm [61]. Microwave is used in sintering [62] as well as in post-processing of parts [63] because it saves time and energy for processing.

While other radiation such as laser is absorbed on the surface of a powder giving rise to surface absorption, microwave gives rise to volumetric absorption. Due to the non-linear effect of the absorption, the temperature of powders increases drastically after some time giving rise to a high temperature at low input microwave energy. This high rise in the temperature, which is called thermal-runaway effect, makes microwave a potential thermal source for high temperature processing. Though it has unpredictable behaviour at different conditions, which needs to be controlled.

In a microwave assisted selective laser melting, microwave is studied as a complementary heat source to l-beam so in its presence, low laser power scanning will take an advantage of the thermal-runaway effect and will not need a high laser power for scanning. This will help to process a higher melting point ceramic, which otherwise furnishes a cracked part [64]. Application of electric field across PB is also useful to decrease sintering time and temperature of ceramics [65].

Application of microwave in PBF as a primary heat source requires it to be concentrated at a small area so its effect can be realized at the desired zone of a processing area without damaging other parts getting built on PB. Application of a microwave localized heating (LMH) applicator has demonstrated the possibility of the localized melting of metal powders. Using the applicator, microwave of wavelength ~12 cm (2.45 GHz) can be concentrated to a diameter of ~1 mm, which is in contrast to laser for which spot size lower than the wavelength is limited by diffraction.

6.3   High Speed Sintering

HSS is a non-beam PBF in which each layer is scanned twice. First, a shape is defined on a layer by ink jetting using a printhead (Fig. 18a) and then, the whole layer is irradiated by a thermal lamp to consolidate that shape (Fig. 18b).

The ink is a radiation-absorbing material, which causes the area on the layer where the ink is jetted to absorb more thermal radiation than other adjacent areas. The role of the printhead is to mark areas on a layer, which constitute a part. The thermal lamp does not mark any area and thus does not need the details of the CAD file of the part. The lamp is thus free from tool path and covers the whole layer from one end to another end. Therefore, the speed of the lamp is relatively fast, which gives the name 'high speed sintering'.

The speed of HSS is the summation of two steps: (1) time taken by ink jet printhead for jetting on a layer, and (2) time taken by lamp for moving over the layer (scanning the

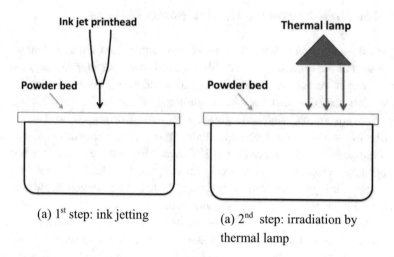

(a) 1st step: ink jetting   (a) 2nd step: irradiation by
                                              thermal lamp

Fig. 18 Schematic diagram of high speed sintering

layer). As scanning by laser in SLS is slow while jetting by a printhead is fast, the speed
is faster in HSS than in SLS. Though for processing a layer in SLS, only one step (laser
scanning) is required while for the same in HSS, two steps (jetting plus lamp scanning)
are required. Since time taken by lamp scanning is insignificant in comparison to the time
taken by jetting, the speed of the process is mostly influenced by the speed of the jetting.

However, if the size of a part is small, HSS will no longer be having advantage over
SLS in terms of speed. It is because there will not be a significant difference between the
time taken by laser scanning and the time taken by ink jet printing (IJP) while the time
taken by lamp scanning will not be insignificant. On the contrary, if the size of a part is
big or if the combined size of many small parts is big so processing to complete a layer
will take longer time, then HSS will be having advantage over SLS in terms of speed,
which will justify 'speed' of high speed sintering.

In a variant of HSS named multi jet fusion (MJF), two types of inks are used—one for
enhancing absorption of the radiation (same as that used in HSS) while the second ink
is an inhibiter type ink used at boundaries of a scanned pattern. The role of the second
ink is to decrease the diffusion of heat outside the boundary. If heat is diffused outside,
it will cause some powder particles to join outside the boundary resulting in a rough side
surface. Using the second ink will make side surface smooth and increase the definition
of a scanned pattern [40]. Use of two types of inks is the same as two types of parameters
in beam PBF—one for scanning at the boundary and the other at non-boundary area [66].

The radiation-absorbing material is carbon black. The heat absorption by carbon must
be enough to join nearby powders. Though, this absorbed heat is not high but sufficient
enough to join low melting point materials such as polyamide and elastomers. This heat

will not be enough to join high melting point materials such as iron or titanium. The process, thus, has limited applications.

Use of carbon especially in the form of graphite is not new to enhance absorption of laser [67]. Presence of carbon black decreases mechanical properties of polyamide in another process (SLS) [68], or its presence does not have a positive influence on mechanical properties [69]. Carbon in the form of a fibre increases both light absorption and mechanical properties [70, 71] but HSS does not use carbon fibre. This brings limitations to properties that can be achieved with HSS.

6.3.1 Energy-Efficiency

HSS does not need a laser and is therefore free from expensive investment. Laser is itself not an efficient energy-conversion device, approximately only from 5 to 70% of electrical energy is converted into laser energy [54, 55]. Most energy-efficient laser (of more than 40% efficiency) is diode laser but it has no widespread use in AM [61, 72] due to its poor beam quality [73]. Since HSS uses a thermal lamp that is almost 98% efficient (~2% is lost in non-thermal light energy), it is more energy-efficient than LPBF as far as the selection of heat source is concerned. In HSS, all powders of the layer get processed, majority of them do not end up in a part. In SLS, only those powders of the layer that end up in a part get irradiated.

Thus in HSS, majority of the powders that do not end up in a part get still processed and directly affected by thermal radiation, which may cause them to be degraded and to lose their ability to be recycled. It is argued that print speed is high and therefore exposure time is low and there will be recyclability but low exposure time is not better than no exposure time. In SLS, there is no such process-induced degradation, risk, and exposure to powders. Besides, there is no such emergency to speed up the process at the cost of non-selection of powders in a layer. It does not mean powders are not at all degraded in SLS, powders are degraded but not at such risk of getting as degraded as in HSS.

Thus, SLS is more efficient than HSS in non-degradation of powders and, therefore, HSS is more powder-inefficient than SLS. Powder-inefficiency of HSS will increase if HSS system becomes bigger to speed up the production process and production, because the size of a layer (PB) will increase and therefore more powders will be at a risk of degradation. Since production of powder is an energy-intensive process [74], HSS being powder-inefficient cannot be energy-efficient as far as utilization of powder is concerned. In this regard, SLS is more energy-efficient. If SLS is not considered energy-efficient, it is not because the process itself is inefficient but because the process uses an energy-inefficient device (laser).

7   Linewise, Pointwise, and Areawise Scanning

7.1   Linewise

In HSS, when a thermal lamp starts to scan from one end of a layer, in the first few moment, it irradiates the whole width but a fraction of the whole length of the layer. Thus, the lamp does not scan any area but just a line on the layer. After few more moments, it covers some more length, then it has covered some area. It starts from scanning a line, and the line moves in a direction normal to it. The line becomes wider with the progress of scanning, which is an area. This is a linewise scanning.

Linewise scanning is fast because it starts from a line. It would not be fast if it had to create a line from scratch by adding many fragments or points. It is free from such troubles. It does not constitute a line by controlling many points, which is reserved for another scanning (pointwise scanning).

Examples of linewise scanning are: stacking of l-beam emitters in a bar across PB and processing it without deflecting beams but switching on/off some emitters and moving the bar [72, 75], or movement of an infrared lamp across PB in MJF [40]. HSS is an example of double linewise scanning—the first linewise scanning takes place when an array of nozzles deposits material on PB while the second takes place when an infrared lamp moves.

7.2   Pointwise

If scanning starts from a point, it is not a linewise but a pointwise scanning, which has to make many overlapping points to make a single line. Thus it will not be as fast as a linewise scanning, but it controls points and has ability to add points in a fashion it wants to add. Thus, it can make a line of any size in any direction, a combination of many lines of various sizes. If the linewise scanning is for fast scanning, it is for fine scanning. If the linewise is to provide high production rate, it is to provide high complexity. If the linewise can make both a line and an area, it can make all three: a point, a line, and an area.

Its examples are scanning using a point source such an e-beam, an l-beam or a jet from a single nozzle. In EPBF, the diameter of an e-beam is the size of a point source; in LPBF, the laser spot size of the l-beam is the size of a point source; and in BJ3DP, the size of the drop on the surface from a nozzle is the size of a point source.

7.3 Areawise

Linewise scanning starts from a line and ends up making an area while pointwise scanning starts from a point and ends up making an area. If they do not make an area, a part will not be made. How they start is important but whether they are able to end up in an area is more important. If the area is important, why not the whole area be scanned at a time.

Considering a thermal lamp that does not follow linewise scanning by starting from one end of a layer and reaching the other end, but faces the whole area and then scans instantly. It means there is a lamp situated over PB, which is switched on and then off. This will cause the whole area to be scanned without any physical movement of the lamp. This instant scanning is faster than linewise scanning. But it has no power of judgement as it can not identify that area that needs to be consolidated. If all scanned area is solidified, there will always be a rectangular block and not a part as per design.

Areawise scanning needs help for making a useful part. It can have help if a thermal radiation coming from the lamp is prevented from reaching every part of an area. This can be done by using a physical mask between the lamp and PB so the mask will prevent the radiation from reaching PB. Therefore, though the whole area is scanned, the effect of radiation will not be realized below the masked area [76].

It can also have help if instead of creating a physical mask, a mask type mechanism is created within lamp so when the lamp is switched on, the radiation does not emanate from the whole area of the lamp but only from its unmasked area [77]. If the radiation comes from the selected area, it will affect only the selected area of PB. Thus, the selected area will be solidified, leaving the remaining area unsolidified, and a part as per design will be formed.

It can again have help if instead of using any type of masks, powders of the selected area of PB is changed so the selected area acts differently when the whole area is irradiated. If a selected area is mixed with a radiation absorbing material (same in HSS), that area will be solidified. If mixing is done as per a design, solidification will happen as per the design, which will help make parts as per design [78]. If the selected area has higher melting point material, that area will not be solidified but other areas will be solidified, which will again help make parts as per design [7].

An example of the areawise scanning is MAPS where microheater array is set above PB. Areawise scanning has been called a layerwise scanning [4]. But the layerwise scanning implies layer by layer scanning and has been used as such for long [79]. It will be thus difficult to distinguish between two meanings of the layerwise scanning: (1) scanning a layer, completing it, going to the next layer and again scanning, and (2) scanning every point of a layer at a time. The name areawise scanning unlike the name layerwise scanning has no such precedence and is free from ambiguity.

7.4 Basic Differences

If the area of PB is big and the exposure area of a lamp is small, the whole area of PB will not be scanned by just switching the lamp on and off. The lamp needs to move to cover the whole area. If the area of PB is big and the exposure area of a lamp is very small, the lamp needs to move not only in one direction but also in other directions to cover the whole area. Though, the lamp was used for areawise scanning when it was still, but if it moves as a point source in different directions, it gives rise to a pointwise scanning.

This brings questions—how small is very small so an area will become a point, which point is not having an area. What if the point is in the form of a line, which line has no area, etc. If the exposure area of the lamp is very small, should the area be called an area or a point? Is there anything more than a physical dimension (point, line, or area) that sets three types of scanning apart?

The purpose of these scannings is to make an area. In a pointwise scanning, the area is made when a point moves in both x- and y-directions in an xy plane. If the point moves only in one direction, it will end up making a line instead of an area. If the point does not move at all, it will not even make a line, there will be just a point. In a linewise scanning, the line needs to move in only one direction (normal to the line) [80] to make an area. If a line does not move, it will not make an area but there will be just a line at the end of the movement. If the line moves in a plane just like a point, i.e. moves in both x- and y-directions, it will make a bigger area. But for making just an area, movement in one direction is sufficient. In an areawise scanning, without any movement of a scanner or lamp, an area can be made.

Hence, in a pointwise scanning, there is a minimum requirement of movement in two dimensions to scan an area. While in a linewise scanning, there is a minimum requirement of movement in one direction to scan an area. In an areawise scanning, there is no requirement of any movement. Though, it can be stated that in an areawise scanning, there is a minimum requirement of movement in zero direction or dimension to scan an area. These are the basic differences among three types of scanning.

Thus, if an area of $500 \ \mu m \times 500 \ \mu m$ is scanned by exposing the area once with a point source of area of $500 \ \mu m \times 500 \ \mu m$, the scanning is an areawise scanning rather than a pointwise scanning. In another example, if an area of $5 \ m \times 5 \ m$ is scanned by moving a lamp of area of $1 \ m \times 5 \ m$ five times in sequence, this type of scanning is a linewise scanning rather than an areawise scanning. Figure 19 shows a schematic diagram of three types of scanning. In a pointwise scanning, an l-beam is used to scan PB (Fig. 19a); in a linewise scanning, a number of l-beam emitters are fixed in a line (Fig. 19b) while in an areawise scanning, an array of heaters covers the whole PB (Fig. 19c).

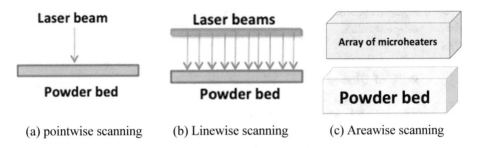

(a) pointwise scanning   (b) Linewise scanning   (c) Areawise scanning

Fig. 19 Schematic diagrams of pointwise, linewise, and areawise scanning

8   Binder Jet Three Dimensional Printing

BJ3DP is a PB non-fusion process in which binder from an ink jet printhead is used to join powders (Fig. 20). The process is similar to beam PBP except the job of a beam is done by a binder jet. It is also called 3D printing, but this name can be confused with general 3D printing, a synonym for AM [81], and therefore, BJ3DP is used instead.

Binder does not melt powders, or powders are joined without going through any type of melting. Hence, different types of powders irrespective of their melting points can be processed. Since a binder is usually not the constituent of a final part, it needs to be separated from the final part without disturbing the shape and dimension of the part. This requires post-processing with an aim to fulfil: (1) removal of the binder so that a part will not continue to remain impure with it, (2) solid state sintering so the structure of the part does not change, (3) infiltration so density and strength is enhanced. Requirement of

Fig. 20 Schematic diagram of binder jet three dimensional printing

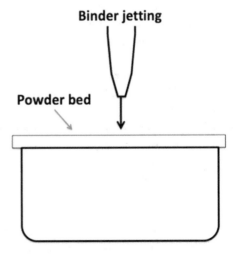

post-processing means BJ3DP is inferior to those AM processes that give final products without post-processing.

8.1   Role of Binder

Binder is jetted on PB, jetting implies the binder is either liquid or air. There is no air based binder, though there is air deposition based process named aerosol jetting. Thus, all jetted materials are liquid. Binder comes from a nozzle of an ink jet printhead—whatever jetted from the nozzle is supposed to act as a binder (to join) or to act as a partial binder (to cause to join). Binder means it is sufficient to bind powders without seeking an active contribution from the powder itself, e.g. joining of iron or copper powders by an organic binder.

The binder joins iron powders as much efficiently as it joins copper powders. If there is a difference in the joining strength, it is due to the difference in the size or surface texture of the powder, not due to the different role of iron as a material than copper. Thus, the powder is neutral and does not contribute to increase the capability of binders. Hence, the situation will remain same if iron or copper is replaced with aluminum or tungsten.

What will happen if powder is not neutral. Then, its constituents might be different. Possibly, it is made of binders, or is itself a binder, or is admixtured with a few percentages of binder, or plays different roles, and one of its roles is to act as a binder in the presence of certain materials. In these cases, whatever be jetted from a nozzle may not be completely a binder. But it is a partial binder, because if it does not trigger reaction and cause binding to happen on PB, the nozzle will not be able to make a pattern and help create a 3D shape. An example of a partial binder is using water as a binder on cement powders. Water reacts with cement and binds cement powders, the same water may not be used as a binder for iron powders.

A partial binder is used if a binder:

(1) Solely jetted from a nozzle is not able to go through the thickness of a layer, some other binder present in the layer is required to make up the binding process,
(2) Clogs a nozzle,
(3) Does not improve the rheology of a liquid jet, or help form better drops,
(4) Leaves residue after post-processing,
(5) Is not required, e.g. water as a partial binder on cement or plaster,
(6) Is not intended to be removed during post-processing but is required as the constituent of a final part.

The aim of the jetted material is to bind those areas of PB where it is jetted. But if it does not bind only those areas, then an object that is the negative 3D image of a 3D object is formed. This process is called selective inhibition sintering (SIS), implying the jetted material inhibits binding [7]. This is possible if the jetted area has lower ability to bind while the other area is full of binder so during post-processing heat treatment, the other area, unlike the jetted area, binds as the jetted area is immune to the high temperature.

The main difference between BJ3DP and SIS:

(1) If a big part is to be made, then in BJ3DP, it requires a lot of binder to be jetted while in SIS, it requires to be jetted only on boundaries. Thus, more binder is saved using SIS.
(2) If a complex part is to be made, then in BJ3DP, the part bound with binder (a green part) is taken to a furnace for post-processing while in SIS, whole powder container containing jetted area and surrounding powder is taken. In most geometries, surrounding powder only on one side of the jetted line or area is intended to become a part while that of the other side becomes waste. Thus, in SIS, the powder that does not contribute to the final part is also subjected to furnace treatment and becomes unusable. Thus, more powder is saved using BJ3DP.

Figure 21 shows the difference between BJ3DP and SIS. Figure 21a shows binder deposition in BJ3DP, binder is deposited in yellow area, the part forms (Fig. 21b) as per deposited area. Figure 21c shows inhibiter deposition in yellow area, the part forms (Fig. 21d) from other area that is surrounding powder (red in color). The part consists of two components: a rectangular area outside the yellow area and a square inside the yellow area. The rectangular component may not be used and becomes waste.

9 Photopolymer Bed Process

Photopolymer bed process (PPBP) is SL [82], microstereolithography [83], digital light processing [84], large area maskless photopolymerization [85], etc. The name 'PPBP' has some edge over the name 'SL' and other names [86, 87].

In PPBP, photo sensitive polymers are irradiated with light (UV or other lasers), which causes photons to interact with electrons and polymer molecules. The interaction creates either free radicals by breaking bonds or ions (cations) by removing electrons. The free radicals or cations bond with other polymer molecules, which enables them to bond again with other polymers resulting in an increase in a chain length, viscosity, gelling, and molecular weight.

Polymer chain formation is known as polymerization. Since low-energy photon is sufficient to trigger polymerization, laser power of few mW is sufficient, which is in contrast to high laser power of hundred watts used in metal powder based AM as high-energy photons are required to break metallic bonds and melt them.

With photopolymer, the process can work in many orientations. This could not be possible if the photopolymer were solid particles. In the majority of cases, the process

Fig. 21 Role of binder and inhibiter in binder jet three dimensional printing and selective inhibition sintering

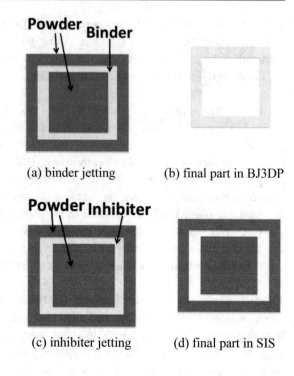

(a) binder jetting   (b) final part in BJ3DP

(c) inhibiter jetting   (d) final part in SIS

works in a general orientation (Fig. 22a) where l-beam is irradiated on the photopolymer from the top and the solidified part moves down inside the liquid. This orientation allows to make big parts or many parts—it can lead to mass production [28].

But, what if the orientation is turned upside down? The approach to the substrate will also be turned upside down. The l-beam needs to approach the substrate by being present not above but below the substrate. The photopolymer will be required to be constrained from spilling. The substrate will have no room to go down but to go up and, therefore, the solidified part has to move up layer after layer. In this orientation (Fig. 22b), the l-beam passes through a glass window and solidifies the liquid present between the substrate and the glass. To prevent the solidified material from being attached on the glass, a film is sealed on the glass. Successive layers are formed by moving the substrate up and filling up the gap between the solidified layer and the glass by recoating.

This orientation may not allow to make a big part as made in the general orientation but facilitates to make a part with minimum supply of liquid enough to fill the gap between the window and the on-going solidified part [88]. Other orientations are also possible such as sidewise orientation (Fig. 22c) [89]. Different orientations allow parts to grow in different orientations (Fig. 22d–f) [84].

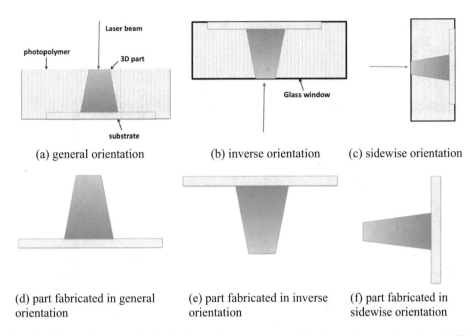

Fig. 22   Three orientations in photopolymer bed process (**a–c**) and the parts made therefrom (**d–f**)

10   Slurry Bed Process

Slurry is a mixture of solid and liquid in which the solid is suspended in the liquid. To suspend solid particles, dispersant is required, which gets adsorbed on the surface of solid particles and does not let them come in contact with each other to agglomerate. Increasing more dispersant is counterproductive as it will decrease dispersion and increase agglomeration by bonding solid particles with dispersant polymer chains.

The liquid medium is called solvent that may be water or other polymeric solutions (ethanol, acetic acid, etc.); water is not suitable for some materials that get oxidized or solidified in reaction with it [90]. Slurry requires plasticizer to increase its fluidity. Besides, it requires binder to hold particles together when dried.

Slurry bed process has following advantages over PBP:

1. It is not convenient to make a layer using sub-micron sized powders as they repel each other due to electrostatic forces. Using slurry made from them helps make a thin layer not achievable in PBP.
2. PBP does not work well with powders having: broad size distribution, non-spherical shape, or high differences in density as spreading these powders is affected by their properties while spreading a slurry made by them is not affected by their properties.

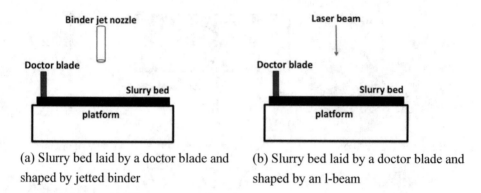

(a) Slurry bed laid by a doctor blade and shaped by jetted binder

(b) Slurry bed laid by a doctor blade and shaped by an l-beam

Fig. 23 Shaping in slurry bed process due to: **a** binder jetting, **b** scanning by an l-beam

3. Powder layer possesses lower density. Using slurry instead will increase the layer density. Initial higher density of the layer may help achieve higher part density.

In this process, with the help of a doctor blade, slurry is laid on a bed, which is then dried so the deposition of the next layer should not deform it (Fig. 23). For an aqueous based slurry, drying can happen by raising the substrate temperature to more than the boiling point of water.

Without increasing the temperature, a layer can also dry by capillary action when the solvent is absorbed by previous layers. The dried layer is selectively treated by an l-beam [91] or a binder jet [92] to create a pattern on it. The processed part, known as a green part, goes through post-processing to improve its properties [93].

For smooth spreading of a layer, viscosity needs to be small. But, if it is small, there will be a higher liquid content in slurry, which needs to be removed during drying—this may give rise to cracks in the slurry. Low viscosity implies there is no high content of solid particles, which gives rise to low density of the layer and then of the part. Low density means low strength that may not be desired unless a porous part is needed for some applications.

10.1   Binding Methods

A dried slurry layer unlike a powder layer has particles bonded with each other. This bonding is not meant to stay permanently but to get dissolved during post-processing. The layer needs to be additionally treated in the same way as treated in PBP to make a pattern on it by creating additional bonds among slurry particles at selected areas. These areas being the constituent of a desired part are not meant to be dissolved during post-processing. The additional bond is created by an l-beam, same as in LPBF, when the

Fig. 24 Binding methods in a
slurry bed process

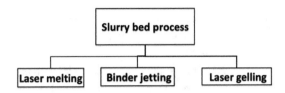

beam partially melts slurry and vaporizes polymers—this process is known as laser slurry deposition [91] or ceramic laser fusion (CLF) [94]. The bond is also created by jetted binder same as in BJ3DP, when the binder fills in pores, locks separate particles and permeates through the slurry layer to bond with an underlying layer—this process is known as laser slurry deposition-print [95].

Bonding can be created by activating binders present in a non-dried slurry layer. Colloidal silica (sol) present in the layer remains dormant unless the layer is scanned by l-beam, the beam converts the sol into gel that bonds adjacent particles. This process is known as selective laser gelling [96, 97]. Thus, there are three binding methods in a slurry bed process—binder jetting (Fig. 23a) laser melting (Fig. 23b) and laser gelling (Fig. 24).

Slurry mixed with photopolymer can be spread in the form of a bed to make a layer. Since the solid content (about 40 vol.%) in photopolymer is low, the process is nearer to photopolymer bed process rather than slurry bed process. This type of slurry is conveniently processed akin to photopolymers. Low solid content means low deflection of l-beam, which allows adequate curing thickness more than the layer thickness, which needs to be obtained to ensure sufficient bonding between two layers. The method of bonding is curing by an l-beam, similar to stereolithography. This process is called lithography based ceramic manufacturing [98].

11 Layerless Powder Bed Process

PBP makes a layer when PB is formed. This layer is of constant thickness because the layer is made by spreading on a planar platform that is positioned parallel to the direction of the spreading. The thickness of the layer will increase when the platform goes down, increasing the separation between the build plane and the platform. Thus, the thickness is changed by changing the separation.

But this changing of the layer thickness does not change the fact there will always be a fixed thickness. Though there will be a number of thicknesses that can be obtained when a part is made—this is how a part property is optimized by combining a number of thicknesses [11]. But for a given layer when the processing is happening, the thickness does not change. This limitation of PBP to not being able to change the thickness unless the processing of a layer is completed gives a uniform thickness of a powder layer.

What will happen if thickness for a particular layer does not remain constant but changes, then the solidified layer that is obtained after processing will not be planar but will be of varied thicknesses. This means there will no longer be one thickness but many thicknesses of a layer, implying how a layer thickness is perceived presently will not be applied in this case.

In order to change the thickness of a layer many times during processing, it is necessary to know how a particular layer thickness is obtained and how it can be changed.

A particular layer thickness is obtained because when a platform goes down, a scraper moves from one side of the platform to the other side, filling a gap with powder on the platform. This movement of the scraper from one side to other side helps a constant thickness to be obtained on the platform. Unless the scraper has completed moving and reaching to the other side, it is not possible any more to decrease the platform position or index. But after the scraper has reached the other side, the platform position is decreased. The next movement of the scraper again will give another layer when the scraper reaches the other side. This is the to and fro movement of the scraper, which gives one layer of one thickness for its to movement and another layer of another thickness for its fro movement.

What if when the scraper starts to move, the platform starts to go down—the platform does not go down fast and the scraper brings enough powder to fill the increasing gap on the platform (Fig. 25a). At the end, there is not a layer thickness of constant size but of increasing size. What will happen if a laser beam or a binder jet follows the scraper when the scraper is moving—the laser beam or the binder does not wait to let the scraper reach the end of the platform to start processing or solidifying, instead the processing follows the scraper. It means there is processing, then there is no processing, then there is processing—the cycle goes on before the scraper reaches the end. Each processing means a fraction of layer is solidified, giving a small solidified bed. Each no processing means a fraction of layer is going to be spread, but at a slightly different height.

This cycle reminds of the usual cycle that happens in PBP when there is processing of a layer followed by no processing during the spreading of the layer—the cycle requires the completion of a layer.

While this cycle happens without the completion of a layer. At the end of this cycle, there will a triangular solid block consisting of many steps (Fig. 25b). The number of steps is equal to the number of times of the processing. Formation of the step means the downward movement of the platform was not too fast to disturb the planarity of the step. At the beginning, the step has the minimum height corresponding to the initial layer thickness while at the end, the step has the maximum height equal to the layer thickness achieved at the end. It is feasible if the processing is possible from minimum to maximum layer thickness. If there is no much difference between the minimum and the maximum thickness, the height of the block will be small.

The shape of the layer thus obtained is not the part of state-of-the-art AM. The layer must be planar so a CAD model is sliced parallelly to give planar layers, and so that the difficulty of processing a layer will not change from one location to another location within a layer due to variations in the thickness.

Fig. 25 Formation of an inclined layer (**b**) by decreasing the platform while moving the scrapper (**a**)

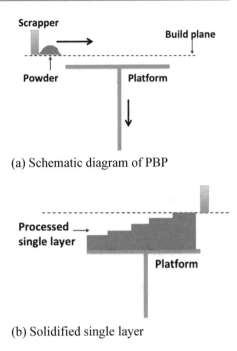

(a) Schematic diagram of PBP

(b) Solidified single layer

When AM is called ALM, the name conveys that the layer is planar. The reason is that ALM is not advanced enough to accommodate all types of layers whether they are non-planar [99–101], curved [102, 103], spiral, or of increasing or decreasing thickness. This is the reason why ALM in the case of powder bed process is not free from 'bed'. It is no exaggeration to say laser based PBP provides a bed, literally, to sleep on before a laser beam comes knocking.

Thus in the state-of-the-art PBP, there is an unwritten pledge between a platform and a scraper that the platform will remain faithful and will not move down unless the scraper completes its journey and reaches the other side of the platform.

Formation of bed in PBP requires cooperation from the platform. The platform provides constant support so a layer is formed, comprising of myriad of powders having unmovable positions. These positions of the powder can move during processing, but they will not move during processing for the reason that the platform fails to provide a support. But when a layer is formed by constantly moving down the platform, it generates a layer having slope, which does not guarantee that the powders will not move and change the position, leading to a possible collapse of the inclined layer.

Thus, processing of a layer after its formation as happens in PBP will not work in an inclined layer. The waiting that happens in PBP that unless the scraper does not reach the other end, processing of a layer will not start will not work in the case of an inclined layer. There needs to fix a layer before the powder of the inclined layer starts to unsettle because the slope starts to become steep with a constant moving down of the platform.

Before the layer becomes steep, the layer is not steep—this means within some tolerance, the inclined layer can be considered a layer resembling a bed. This means, within a fraction of distance from the start of the movement of the scrapper, there is no difference between an inclined layer and a non-inclined layer (if formed). As the processing of the inclined layer within the fraction of the distance does not bring a non-uniformity in the property that could be expected to be obtained when processing a inclined layer. Thus, the inclined layer is not different from a powder bed when is processed not far late from its time of formation.

Consequently, processing of an inclined layer if done within a certain time after it starts to get steep is not different from processing a number of small powder beds arranged in an increasing steps comprising the layer.

11.1   Formation of a Part

Formation of an inclined layer and its processing leads to the creation of a block having steps—this is a processed layer. Since a part consists of many layers—a part must be conceptualized in the form of an aggregation of such layers.

Considering a part which consists of more than one layer. For fabrication of the second layer, processing must happen on the first layer which is a triangular block having one end at the level of the build plane while the other end is at the nearest to the platform surface (Fig. 26). Powder spreading followed by step-wise processing can start at any of the two ends of the platform. At one end, where the upper surface of the block is at the level of the build plane, the platform needs to go down to accommodate the powder and start the processing. But, after that, the platform does not need to go down because there already exists enough space above the sloped surface to be filled up and processed.

Though powder spreading due to scraper is required to level the surface. Processing though happens through small powder bed by small powder bed as happened while make the first layer. But, the processing needs to be done on a surface full of steps, which was not the case while processing the first layer. This requires to overcome extra difficulty in processing. There would not be any such difficulty should there was a planar layer.

Fig. 26 For making second
layer on an inclined layer

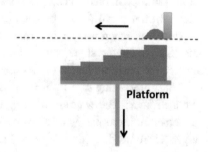

Fig. 27 A circular platform rotating and going down for making spiral parts

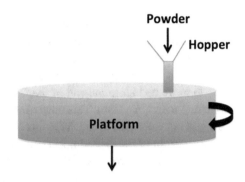

This is the reason why layerwise manufacturing is planar layerwise manufacturing. In the case of a planar layer, when the scraper returns from the other end to the first end while making the next layer, the scraper does not find different thicknesses or gaps to fill. But in the case of inclined layer, the problem starts when the scraper returns to make the next layer as the processed layer is stepwise decreasing. There was no such problem when the first layer was getting processed. During the processing of the first layer, there is no much difference in difficulty in making either planar layer or inclined layer.

What will happen if only the first inclined layer will be formed? This is not possible because a useful part consists of more than one layer. The problem starts when the scrapper returns to make the next layer. Is it possible that the scrapper does not need to return? It is possible when the scrapper moves in a circular path, and with a constant downward movement of the platform, the scrapper acquires a spiral path, making it not to encounter the same point of the path [1].

Thus, the scraper always makes small powder beds. How small the small powder bed is depends upon the speed of the downward movement of the platform. The scraper does not need to move when the platform is not big and can rotate (Fig. 27). It is the relative movement between the platform and the scrapper that gives the required powder bed.

This type of powder spreading and consolidation will lead to a cylindrical part having the transition between two successive layers that will not be as visible as found in layerwise, which is an advantage. If a circular track is big, parts of any shape (not only the cylindricals) can be made by utilising a small section of the track.

11.2   Why It Is Non-layerwise

A spiral path does not have a clear transition from the first track to the second track while the circular path will not be having any transition to the second track unless there is a clear transition. This is the difference between non-layerwise and layerwise bed process. In layerwise (circular track), there comes a point when index moves up from the first track to the second track (Fig. 28a), this point comes after completion of the first track. This index shows a discontinuity in fabrication, which makes it what is layerwise. Without this discontinuity, the ongoing fabrication will not resume.

In non-layerwise (spiral track), there never comes a point for the index to move up from the first track to the second track, but the transition takes place—it means the index starts to move up before the end of the first track and continues moving up after the end of the first track (Fig. 28b). This upward movement of index does not consume a point in the space but consumes a line. This line consists of the end section of the first track and the start section of the second track.

The track due to this curved line is not different from an inclined track due to a linear line made in PBP (Fig. 25) as both tracks consist of steps. The steps of the inclined track give an indication how the index moves up stepwise after the formation of a small powder bed each time.

Thus, spiral tracks can be understood to be a mixture of various lengths of powder bed. During transition, the powder beds are small while at other locations of the track, the powder bed can be long. Depending upon the angle of spiral with vertical direction, the lengths of the powder bed at various locations will vary.

In circular tracks, there is one length of a powder bed—the length of the circular track. Both circular and spiral tracks have one common thing—both make powder beds. This is the reason why both are bed process. But bed means a layer, then why both are not layerwise process?

A layerwise is understood to be those processes, which make products as per the slice of a CAD model, where the slice is a planar layer. This planar layer does not represent the myriad types of powder beds formed in spiral tracks. This planar layer is not meant to represent the composition of a particular track, but whether a particular track is planar or not. As a spiral track is not represented by a planar layer, it does not come under layerwise.

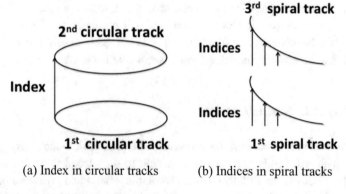

(a) Index in circular tracks   (b) Indices in spiral tracks

Fig. 28 Index in circular (**a**) and spiral (**b**) tracks

But making a spiral part does not necessarily require a CAD model to be sliced in spiral layers. A CAD model sliced in planar layers can also make spiral tracks and then a spiral part. What is required in this case is the incorporation of the downward movement of the platform during the formation of a planar layer—the result will be spiral track. Whether this is layerwise or non-layerwise? It should not be considered layerwise because the spiral track that is formed does not represent the planar layer. Besides, the formation of spiral tracks is due to not following the right set of steps of a process. The right process steps do not include the downward movement of a platform during the formation of a layer. The transformation of planar layers into spiral tracks demonstrate either the realization of another process or a defect in the original process. Though this process defect is not visible because it comes as a manufacturing strategy, showing how the limitation of layerwise can be overcome by manufacturing cleverness.

What if layerwise includes not only the planar layer but spiral layer as well, then slicing a CAD model in spiral layers followed by making a spiral part will not be non-layerwise. What will happen then? Then, there will be a few process left, which will be non-layerwise. But how ALM is understood or will be understood will not change the fact there is a difference between a process using planar layers and that using non-planar layers, which will be required to be taken into account. This justifies the present classification, which takes into account this difference and classifies planar layer process as layerwise and non-planar layer process as non-layerwise.

References

1. Hauser C, Sutcliffe C, Egan M, Fox P (2005) Spiral growth manufacturing (SGM)—a continuous additive manufacturing technology for processing metal powder by selective laser melting. In: SFF symposium proceedings, Texas
2. Dudley K (2015) 3D printing using spiral buildup. US patent US20140265034A1
3. Baumers M, Tuck C, Hague R (2015) Selective heat sintering versus laser sintering: comparison of deposition rate, process energy consumption and cost performance. In: SFF proceedings, pp 109–121
4. Holt N, Horn AV, Montazeri M, Zhou W (2018) Microheater array powder sintering: a novel additive manufacturing process. J Manuf Process 31:536–551
5. Brown R, Morgan CT, Majewski CE (2018) Not just nylon—improving the range of materials for high speed sintering. In: SFF proceedings, pp 1487–1498
6. Jerby E, Meir Y, Salzberg A et al (2015) Incremental metal-powder solidification by localized microwave-heating and its potential for additive manufacturing. Addit Manuf 6:53–66
7. Khoshnevis B, Zhang J, Fateri M, Xiao Z (2014) Ceramics 3D printing by selective inhibition sintering. In: SFF proceedings, pp 163–169
8. Thomas HR, Hopkinson N, Erasenthiran P (2006) High speed sintering—continuing research into a new rapid manufacturing process. In: SFF proceedings, pp 682–691
9. Liravi F, Vlasea M (2018) Powder bed binder jetting additive manufacturing of silicone structures. Addit Manuf 21:112–124

10. Kumar S (2022) Role of post-process. In: Additive manufacturing solutions. Springer, Cham, pp 41–56
11. Kumar S (2014) Selective laser sintering/melting. In: Comprehensive materials processing, vol 10. Elsevier Ltd, pp 93–134
12. Kumar S (2022) Advantage. In: Additive manufacturing solutions. Springer, Cham, pp 7–29
13. Kumar S (2020) Beam based solid deposition process. In: Additive manufacturing processes. Springer, Cham, pp 93–109
14. Onuike B, Bandyopadhyay A (2019) Additive manufacturing in repair: influence of processing parameters on properties of Inconel 718. Mater Lett 252:256–259
15. Li W, Yang K, Yin S et al (2018) Solid-state additive manufacturing and repairing by cold spraying: a review. J Mater Sci Technol 34(3):440–457
16. Huang CJ, Wu HJ, Xie YC et al (2019) Advanced brass-based composites via cold-spray additive-manufacturing and its potential in component repairing. Surf Coat Technol 371:211–223
17. Priarone PC, Campatelli G, Catalano AR, Baffa F (2021) Life-cycle energy and carbon saving potential of wire arc additive manufacturing for the repair of mold inserts. CIRP J Manuf Sci Technol 35:943–958
18. Griffiths RJ, Petersen DT, Garcia D, Yu HZ (2019) Additive friction stir-enabled solid-state additive manufacturing for the repair of 7075 aluminum alloy. Appl Sci 9(17):3486
19. Kumar S (2022) Application. In: Additive manufacturing solutions. Springer, Cham, pp 93–110
20. Aziz NA, Adnan NAM, Wahab DA, Azman AH (2021) Component design optimisation based on artificial intelligence in support of additive manufacturing repair and restoration: current status and future outlook for remanufacturing. J Clean Prod 296:126401
21. Thompson MK, Moroni G, Vaneker T et al (2016) Design for additive manufacturing: trends, opportunities, considerations, and constraints. CIRP Ann 65(2):737–760
22. Shi X, Ma S, Liu C et al (2016) Performance of high layer thickness in selective laser melting of Ti6Al4V. Materials 9
23. Shamsdini S, Shakerin S, Hadadzadeh A et al (2020) A trade-off between powder layer thickness and mechanical properties in additively manufactured maraging steels. Mater Sci Eng A 776:139041
24. Leicht A, Fischer M, Klement U et al (2021) Increasing the productivity of laser powder bed fusion for stainless steel 316L through increased layer thickness. J Mater Eng Perform 30:575–584
25. Gao W, Zhang Y, Ramanujan D et al (2015) The status, challenges, and future of additive manufacturing in engineering. Comput Aided Des 69:65–89
26. Lim S, Buswell RA, Le TT et al (2012) Developments in construction-scale additive manufacturing processes. Autom Constr 21:262–268
27. Yu KS, Cheng CW, Lee AC et al (2022) Additive manufacturing of NdFeB magnets by synchronized three-beam laser powder bed fusion. Opt Laser Technol 146:107604
28. Kumar S (2022) Mass production. In: Additive manufacturing solutions. Springer, Cham, pp 145–167
29. Wong H, Dawson K, Ravi GA et al (2019) Multi-laser powder bed fusion benchmarking—initial trials with Inconel 625. Int J Adv Manuf Technol 105:2891–2906
30. Kniepkamp M, Harbig J, Seyfert C, Abele E (2018) Towards high build rates: combining different layer thicknesses within one part is selective laser melting. In: SFF symposium proceedings, pp 2286–2296

31. Schwerz C, Schulz F, Natesan E, Nyborg L (2022) Increasing productivity of laser powder bed fusion manufactured Hastelloy X through modification of process parameters. J Manuf Process 78:231–241

32. Pauzon C, Forêt P, Hryha E et al (2022) Argon-helium mixtures as laser-powder bed fusion atmospheres: towards increased build rate of Ti-6Al-4V. J Mater Process Technol 279:116555

33. Sow MC, De Terris T, Castelnau O et al (2020) Influence of beam diameter on laser powder bed fusion (L-PBF) process. Addit Manuf 36:101532

34. Zhang Y, Bernard A, Harik R et al (2017) Build orientation optimization for multi-part production in additive manufacturing. J Intell Manuf 28(6):1393–1407

35. Calignano F (2014) Design optimization of supports for overhanging structures in aluminum and titanium alloys by selective laser melting. Mater Des 64:203–213

36. Barba D, Alabort C, Tang YT et al (2020) On the size and orientation effect in additive manufactured Ti-6Al-4V. Mater Des 186:108235

37. Gong X, Lydon J, Cooper K, Chou K (2014) Beam speed effects on Ti–6Al–4V microstructures in electron beam additive manufacturing. J Mater Res 29(17):1951–1959

38. Tang M (2017) Inclusions, porosity and fatigue of AlSi10Mg parts produced by selective laser melting. PhD thesis, Carnegie Mellon University

39. Sun Z, Tan X, Tor SB, Yeong WY (2016) Selective laser melting of stainless steel 316L with low porosity and high build rates. Mater Des 104:197–204

40. Sillani F, Kleijnen RG, Vetterli M et al (2019) Selective laser sintering and multi jet fusion: process-induced modification of the raw materials and analyses of parts performance. Addit Manuf 27:32–41

41. Kahnert M, Lutzmann S, Zaeh MF (2007) Layer formations in electron beam sintering. In: SFF symposium proceedings, pp 88–99

42. Sun SH, Hagihara K, Ishimoto T et al (2021) Comparison of microstructure, crystallographic texture, and mechanical properties in Ti–15Mo–5Zr–3Al alloys fabricated via electron and laser beam powder bed fusion technologies. Addit Manuf 47:102329

43. Edinger R (2018) Laser heated electron beam gun optimization to improve additive manufacturing. In: SFF symposium proceedings, pp 2297–2304

44. Bakeev IY, Klimov AS, Oks EM, Zenin AA (2018) Generation of high-power-density electron beams by a forevacuum-pressure plasma-cathode electron source. Plasma Sources Sci Technol 27:075002

45. Lee HJ, Ahn DG, Song JG et al (2017) Fabrication of beads using a plasma electron beam and Stellite21 powders for additive manufacturing. Int J Precis Eng Manuf Green Technol 4:453

46. Sigl M, Lutzmann S, Zaeh MF (2006) Transient physical effects in electron beam sintering. In: SFF symposium proceedings, pp 464–477

47. Krumeich F. Properties of electrons, their interactions with matter and applications in electron microscopy. Laboratory of Inorganic Chemistry, ETH Zurich

48. Azhirnian A, Svensson D (2017) Modeling and analysis of aberrations in electron beam melting (EBM) systems. Master thesis, Chalmers University of Technology

49. Korner C (2016) Additive manufacturing of metallic components by selective electron beam melting—a review. Int Mater Rev 61(5):361–377

50. Yan W, Ma W, Shen Y (2020) Powder sintering mechanisms during the pre-heating procedure of electron beam additive manufacturing. Mater Today Commun 25:101579

51. Cordero ZC, Meyer HM III, Nandwana P, Dehoff RR (2017) Powder bed charging during electron-beam additive manufacturing. Acta Mater 124:437–445

52. DebRoy T, Wei HL, Zuback JS et al (2018) Additive manufacturing of metallic components—process, structure and properties. Prog Mater Sci 92:112–224

53. Townsend A, Senin N, Blunt L et al (2016) Surface texture metrology for metal additive manufacturing: a review. Precis Eng 46:34–47
54. Hecht J (2018) Understanding lasers: an entry-level guide. Wiley
55. Li L (2000) The advances and characteristics of high-power diode laser materials processing. Opt Lasers Eng 34(4–6):231–253
56. Bidare P, Bitharas I, Ward RM et al (2018) Laser powder fusion in high-pressure atmospheres. Int J Adv Manuf Technol 99:543–555
57. Montgomery C, Farnin C, Mellos G et al (2018) Effect of shield gas on surface finish of laser powder bed produced parts. In: SFF symposium proceedings, pp 438–444
58. Ferrar B, Mullen L, Jones E et al (2012) Gas flow effects on selective laser melting (SLM) manufacturing performance. J Mater Process Technol 212:355–364
59. Steen WM, Majumder J (2010) Laser material processing. Springer-Verlag London Limited
60. Enneti RK, Prough KC, Wolfe TA et al (2018) Sintering of WC-12%Co processed by binder jet 3D printing (BJ3DP) technology. Int J Refract Met Hard Mater 71:28–35
61. Pinkerton AJ (2016) Lasers in additive manufacturing. Opt Laser Technol 78A:25–32
62. Leonelli C, Veronesi P, Denti L et al (2008) Microwave assisted sintering of green metal parts. J Mater Process Technol 205(1–3):489–496
63. Salehi M, Maleksaeedi S, Nai MLS, Gupta M (2019) Towards additive manufacturing of magnesium alloys through integration of binderless 3D printing and rapid microwave sintering. Addit Manuf 29:100790
64. Buls S, Vleugels J, Hooreweder BV (2018) Microwave assisted selective laser melting of technical ceramics. In: SFF proceedings, pp 2349–2357
65. Hagen D, Kovar D, Beaman JJ (2018) Effects of electric field on selective laser sintering of yttria-stabilized zirconia ceramic powder. In: SFF symposium proceedings, pp 909–913
66. Tian Y, Tomus D, Rometsch P, Wu X (2017) Influences of processing parameters on surface roughness of Hastelloy X produced by selective laser melting. Addit Manuf 13:103–112
67. Wagner T, Hofer T, Knies S et al (2005) Laser sintering of high temperature resistant polymers with carbon black additives. Int Polym Process 19(4):395–401
68. Athreya SR, Kalaitzidou K, Das S (2010) Processing and characterization of a carbon black-filled electrically conductive Nylon-12 nanocomposite produced by selective laser sintering. Mater Sci Eng A 527(10–11):2637–2642
69. Hong R, Zhao Z, Leng J et al (2019) Two-step approach based on selective laser sintering for high performance carbon black/polyamide 12 composite with 3D segregated conductive network. Compos Part B Eng 176:107214
70. Goodridge RD, Shofner ML, Hague RJM et al (2011) Processing of a polyamide-12/carbon nanofibre composite by laser sintering. Polym Test 30(1):94–100
71. Chatham CA, Long TE, Williams CB (2019) A review of the process physics and material screening methods for polymer powder bed fusion additive manufacturing. Prog Polym Sci 93:68–95
72. Arredondo MZ, Boone N, Willmott J et al (2017) Laser diode area melting for high speed additive manufacturing of metallic components. Mater Des 117:305–315
73. Santos ES, Shiomi M, Osakada K, Laoui T (2006) Rapid manufacturing of metal components by laser forming. Int J Mach Tools Manuf 46(12–13):1459–1468
74. Fredriksson C (2019) Sustainability of metal powder additive manufacturing. Procedia Manuf 33:139–144
75. Dallarosa J, O'Neill W, Sparkes M, Payne A (2016) Multiple beam additive manufacturing. Patent WO2016201309A1
76. Hermann DS, Larson R (2008) Selective mask sintering for rapid production of parts, implemented by digital printing of optical toner masks. In: NIP & digital fabrication conference

77. Farsari M, Claret-Tournier F, Huang S et al (2000) A novel high-accuracy microstereolithography method employing an adaptive electro-optic mask. J Mater Process Technol 107(1–3):167–172

78. Ellis A, Noble CJ, Hopkinson N (2014) High speed sintering: assessing the influence of print density on microstructure and mechanical properties of nylon parts. Addit Manuf 1–4:48–51

79. Deckard CR, Beaman JJ, Darrah JF (1992) Method for selective laser sintering with layerwise cross-scanning. US5155324A

80. Zhu L, Cheng J, Zhou H (2000) Research of rapid prototyping process using linear array of high power laser diodes. In: High power lasers in manufacturing. Proceeding of SPIE 3888

81. Kumar S (2022) Synonym. In: Additive manufacturing solutions. Springer, Cham, pp 1–6

82. Jacobs FP (1992) Rapid prototyping and manufacturing: fundamentals of stereolithography. Society of Manufacturing Engineers, Dearborn, MI

83. Bertsch A, Lorenz H, Renaud P (1999) 3D microfabrication by combining microstereolithography and thick resist UV lithography. Sens Actuators A Phys 73(1–2):14–23

84. Santoliquido O, Colombo P, Ortona A (2019) Additive manufacturing of ceramic components by digital light processing: a comparison between the "bottom-up" and the "top-down" approaches. J Eur Ceram Soc 39(6):2140–2148

85. Rudraraju A, Das S (2009) Digital date processing strategies for large area maskless photopolymerization. In: SFF symposium proceedings, pp 299–307

86. Kumar S (2020) Classification. In: Additive manufacturing processes. Springer, Cham, pp 21–40

87. Kumar S (2020) Liquid based additive layer manufacturing. In: Additive manufacturing processes. Springer, Cham, pp 131–146

88. Chi Z, Yong C, Zhigang Y, Behrokh K (2013) Digital material fabrication using mask-image-projection-based stereolithography. Rapid Prototyp J 19(3):153–165

89. Hafkamp T, Baars GV, Jager BD, Etman P (2017) A trade-off analysis of recoating methods for vat photopolymerization of ceramics. In: SFF symposium proceedings, vol 28, pp 687–711

90. Wu W, Liu W, Jiang J et al (2019) Preparation and performance evaluation of silica gel/tricalcium silicate composite slurry for 3D printing. J Non-Cryst Solids 503–504:334–339

91. Muhler T, Gomes C, Ascheri M et al (2015) Slurry-based powder beds for selective laser sintering of silicate ceramics. J Ceram Sci Technol 06(02):113–118

92. Zocca A, Lima P, Diener S et al (2019) Additive manufacturing of SiSiC by layerwise slurry deposition and binder jetting (LSD-print). J Eur Ceram Soc 39(13):3527–3533

93. Wang HR, Cima MJ, Kernan BD, Sachs EM (2004) Alumina-doped silica gradient-index (GRIN) lenses by slurry-based three-dimensional printing (S-3DP™). J Non-Cryst Solids 349:360–367

94. Tang HH (2002) Direct laser fusing to form ceramic parts. Rapid Prototyp J 8(5):284–289

95. Lima P, Zocca A, Acchar W, Günster J (2018) 3D printing of porcelain by layerwise slurry deposition. J Eur Ceram Soc 38(9):3395–3400

96. Liu FH, Liao YS (2010) Fabrication of inner complex ceramic parts by selective laser gelling. J Eur Ceram Soc 30(16):3283–3289

97. Liu FH, Lee RT, Lin WH, Liao YS (2013) Selective laser sintering of bio-metal scaffold. Procedia CIRP 5:83–87

98. Harrer W, Schwentenwein M, Lube T, Danzer R (2017) Fractography of zirconia-specimens made using additive manufacturing (LCM) technology. J Eur Ceram Soc 37:4331–4338

99. Ahlers D, Wasserfall F, Hendrich N, Zhang J (2019) 3D printing of nonplanar layers for smooth surface generation. In: IEEE 15th international conference on automation science and engineering (CASE), pp 1737–1743

100. Wulle F, Gorke O, Schmidt S et al (2022) Multi-axis 3D printing of gelatin methacryloyl hydrogels on a non-planar surface obtained from magnetic resonance imaging. Addit Manuf 50:102566
101. Ruan J, Tang L, Liou FW, Landers RG (2010) Direct three-dimensional layer metal deposition. J Manuf Sci Eng 132(6)
102. Tan WS, Juhari MAB, Shi Q et al (2020) Development of a new additive manufacturing platform for direct freeform 3D printing of intrinsically curved flexible membranes. Addit Manuf 36:101563
103. Emon OF, Alkadi F, Kiki M, Choi JW (2022) Conformal 3D printing of a polymeric tactile sensor. Addit Manuf Lett 2:100027

Deposition Process

3

1   Classification

On the basis of states of matter, deposition process is divided into four types: solid, liquid, air, and ion (Fig. 1a). This division guarantees that no process will be left out from a classification as every process deals with a feedstock, and every feedstock has a state of matter.

What if a process deals with more than one state of matter? For example, extrusion based deposition works with polymer that has a solid state. Besides, the process works with ink [1] and liquid metal [2], which have liquid states. The process, therefore, will come under both solid and liquid deposition, resulting it not to have a unique place in a classification (Fig. 2a). Thus, using the state of the matter first, and then using the process type does not lead to a unique place. Why not classify the other way round—process type first and then the state of matter. But—how to know a process type? To know it, it needs to know what the deposition mechanism for a deposition process is.

There are six types of deposition mechanisms (melting, extrusion, deformation, jetting, spraying, field force), which classify deposition process into following types: (1) melting based, e.g., LENS [3], WAAM [4], (2) extrusion based, e.g. FDM [5], (3) deformation based, e.g. AFSD [6], (4) jetting based, e.g. ink jet printing (IJP) [7], (5) spraying based, e.g. CSAM [8], and (6) field based, e.g. electrolytic or colloidal solution based AM [9] (Fig. 1b).

What if a process is classified on the basis of a deposition mechanism first and then on the basis of a state of matter. Then there will be no problems related to the classification of extrusion based deposition, but another problem will start. A known process may be working with more than one mechanism, and the process does not have different names due to different mechanisms. For example, IJP works with two mechanisms: jetting and

© The Author(s), under exclusive license to Springer Nature Switzerland AG 2022   95
S. Kumar, *Additive Manufacturing Classification*, Synthesis Lectures on Engineering, Science, and Technology, https://doi.org/10.1007/978-3-031-14220-8_3

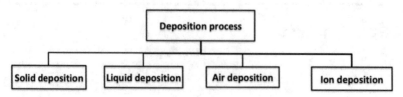

(a) Classification due to states of matter

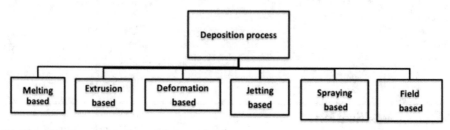

(b) Classification due to deposition mechanisms

Fig. 1 Classification of deposition process on the basis of: **a** state of matter, **b** deposition mechanism

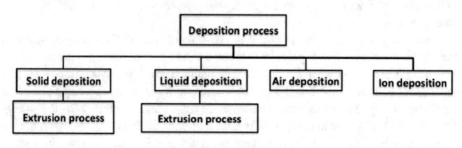

(a) Same process under different states of matter

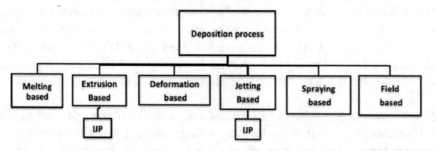

(b) Same process under different mechanisms

Fig. 2 Problems in classification made on the basis of: **a** state of matter, **b** deposition mechanism

extrusion. For both mechanisms, the name of the process is IJP. Therefore, IJP does not have unique place but two places in the classification (Fig. 2b).

Thus, there is a problem in the classification (Fig. 2b), but this problem is not because a process works with two mechanisms when the name of the process is changed. But because there are two processes working with different mechanisms while the name of the two different processes are not different. There would not be any problem if IJP working due to extrusion should have a name IJP-extrusion and IJP working due to jetting should have a name IJP-jetting.

Therefore, the problem in the classification (Fig. 2b) is not because the classification is not able to provide a unique place for a process but because the process used in the classification does not have a unique name. Consequently, if the classification based on mechanism (Fig. 1b) does not provide a unique place for a process, it is because AM is not free from arbitrary process names.

Though there is again a problem in the classification (Fig. 1b). It is based on the mechanism. What if the number of mechanisms increase in future? Then, the classification needs to be modified. It shows that the classification (Fig. 1b) is only for the present time while the classification (Fig. 1a), though having problem, is not restricted only to the present time as the number of states of matter will not increase in future.

2 Deposition Mechanism

The description of the mechanisms is given below:

2.1 Melting

In this type, feedstock is melted. The melting is a precondition for deposition. It comprises of all processes which are beam based such as melting due to laser beam or electron beam or plasma beam, or melting due to an arc. The feedstock can be powder, wire, rod, or solid of any other shapes.

Since melting is possible with only one state of matter, i.e. solid, the classification can work in either way. If a deposition process is first classified on the basis of states of matter and then the mechanism, this will give a unique place for the process. Similarly, if the process is first classified on the basis of the mechanism and then the states of matter, it will also work.

2.2   Extrusion

As per Oxford dictionary, the meaning of extrusion is 'the act of forcing or pushing something out of something'. A material will be forced or pushed out of something (die, nozzle or orifice). If the material is unable to come out on its own, it needs to be plastically deformed to flow out.

Thus, anything (solid metal, solid polymer, viscous liquid) that can be plastically deformed in a nozzle in order to get a shape [10] will fulfil the condition of this mechanism.

In the case of a polymer, it is heated so it acquires deformability and then forced by piston so it flows out of the nozzle. If the material is not pure polymer but a mixture of polymer and some other material, the mixture must have deformability to be extruded. The mixture will get deformability if it has enough amount of polymers. This is the polymer that gives extrudability to a polymer based material. In AM, extrusion is known mainly for its relation with solid polymer because of the widespread use of FDM systems.

Extrusion can be achieved in a metallic alloy by heating it at a higher temperature, and then be forced through a die [11] while it can be achieved in ceramic by mixing with water and then extruding [12]. Thus, extrusion mechanism can be applied for both states of matter, i.e. solid and liquid.

2.3   Deformation

In this type, the feedstock needs to be plastically deformed. If feedstock (wire, rod, powder) is placed on a substrate and is plastically deformed so it will attach to the substrate, a process will come under this mechanism.

Since deformation works with only one state of matter, i.e., solid, the classification can work in either way—first process, then state of matter, or vice versa.

2.4   Comparison Between Deformation and Extrusion

Plastic deformation of a feedstock and then its deposition comes under extrusion. But, the deformation takes place within a nozzle, giving rise a shape to the deformed material.

Deformation based deposition does not rely on deformation before material reaches a substrate. Thus, there is no effort to deform feedstock and give it a shape before it reaches the substrate. While in extrusion based deposition, there is no effort to deform extrudate after it is deposited.

In extrusion based deposition, the plastic deformation is meant to get a shape of the extrudate that in turn gives a 3D structure. If there is no optimum extrusion, it will affect the accuracy of the structure. Getting a well-defined extrudate is the primary aim of this mechanism. The property of the structure is important, but the property is not achieved at the cost of its accuracy. For example, if there is a possibility to obtain an extrudate that will not be able to give a well-defined structure, but will give good properties—this is not preferred in this mechanism.

In deformation based deposition, the plastic deformation (or extrusion on a substrate) is meant to achieve a mechanical property. This is the primary aim of this mechanism. The deformation is increased or decreased depending upon its effect on the dynamic recrystallization and on the grain size, but the aim of the deformation is not to get a particular shape of an extrudate with an aim to fulfil the accuracy of certain 3D structure (as happened in extrusion based deposition). This mechanism can not ignore mechanical properties to get a well-defined structure.

AFSD can not expect to make a delicate structure by deforming as the act of deforming expects certain minimum strength from a previous structure. If the previous structure is delicate, it will yield before further deformation on it will be complete. This requirement means it can not compete with FDM which does not rely on deformation to carry out the building of a structure. FDM relies on laying out extrudate on previous structures to carry forward ongoing build. This act of laying does not demand as much strength from the previous structure as the act of deforming demands.

Though placing an extrudate in FDM may not be free from deformation, and the act of placing may cause deformation on a previous structure and on extrudate itself. It will depend upon the temperature during placing, cooling ability of the previous structure, time of deposition and the material. And this deformation requires to be taken into account for attaining the accuracy. The extent of the deformation may depend upon the case to case basis, but the deformation is not required as much as in AFSD. If there is not even this deformation, there will not be a source of any inaccuracy in FDM due to the deformation. Comparing FDM with AFSD, as far as the source of inaccuracy due to the deformation is concerned, FDM is better than AFSD as AFSD can not avoid deformation.

Though deformation during laying an extrudate in FDM can be a source of inaccuracy, the deformation during the same step in AFSD is a reason for building a structure. It is not an aim in AFSD to build a structure that's accuracy is guaranteed only when no deformation is taking place—the same can not be an aim in FDM.

Though there can be a process which relies on both mechanisms (deformation and extrusion), i.e. deformation within a nozzle to get an extrudate, and deformation of this extrudate on the substrate to get bonding [13].

2.5 Jetting and Spraying

As per Oxford dictionary, jetting means quick expulsion of a stream of liquid or gas from an opening while spraying means sending very small drops of liquid through air.

Spraying requires the help of a gas to send droplets. The gas has no other purpose than to take drops from one point to another point. It is the droplet that has utility at the site. It is different from jetting where a liquid drop can go from one point to another point without the help of a carrier.

Thus, jetting based AM implies that whatever is jetted from a nozzle or printhead becomes part of the ongoing build, while spraying based AM implies that whatever comes out from a nozzle will not necessarily become part of the ongoing build. Thus cold spray based AM means that only particles and not the carrier gas will become constituent of the build.

Spraying differs from jetting as the former implies spreading out of liquid drops or particles—this spreading will not help get deposition in AM. If spraying based process is used not for spreading or diverging but for converging the spray, then spraying in this case will not be different from jetting when the two are compared in terms of the spreading of particles or droplets. Thus, spraying, which sends a stream of droplets, is not doing anything different from jetting, which also sends a stream of droplets, except the former does with the help of a carrier gas while the latter does without it. This is the reason why aerosol jetting or aerosol jet printing which is a aerosol spraying technique uses 'jetting' instead of 'spraying'.

Since cold spray based AM relies on spraying but not on spreading, it will not be wrong if, taking the inspiration from the name of aerosol jetting, the process can instead be 'cold jet based AM'. Similarly, aerosol jetting relies on spraying but not on spreading, it will not be wrong if, taking the inspiration from cold spray based AM, the process can instead be 'aerosol spraying'.

But, whatever be the name or will be the name, there are two mechanisms: one mechanism that does not rely on carrier gas to send droplets while the other that relies on the carrier gas. Consequently, based on these two mechanisms, there will be two deposition based AM.

Jetting based deposition, such as IJP depends on the quick expulsion of droplets. There are methods to accomplish the quick expulsion—it can be by creating a sudden pressure or temperature in a printhead so the droplets will not be able to hold themselves at an orifice and will leave as per impulse on it.

The quick expulsion is necessary as it provides a droplet an ability to strike a target and not drift away from the intended direction. This ensures repeatability in action by droplets as they will have sufficient momentum to furnish repeatable motion. Thus quick expulsion will provide precision along with a high fabrication speed.

What if the expulsion is not quick because quick expulsion is not required? The precision of deposition is not required? Considering a liquid in a syringe, which is expelled dropwise. A pressure on the liquid is required so the liquid will leave syringe opening drop by drop. This pressure is responsible for creating a stream of drops. But this drop, which is part of such a high speed stream, is not required, and therefore, such pressure is not required. But, a drop is required. A drop will not leave syringe on its own, and therefore, a minimum pressure is necessary so the drop will detach from the syringe opening. This type of minimum pressure does not create a condition that comes under jetting.

Thus, without jetting, liquid drop reaches the desired site. This movement of drops also does not constitute an action that is called spraying. Thus, there is a nozzle which drops a liquid, and a product is formed—this process will not come under any deposition mechanism as the mentioned mechanisms are not sufficient to account for it. A new deposition mechanism is required. The name of the mechanism will be moving or falling. These names are not accustomed as jetting, spraying, or extrusion, but the accustomed names are not able to include moving or falling.

There can be another example. In civil engineering for constructing the wall of a building, a slurry made up of sand, cement and water is deposited. The aim is to let the mixture fall through a nozzle when required. This falling does not come under extruding as there is no pressure applied to shape the mixture. Since, shaping the mixture and making an extrudate is not required. What is required is to open a valve fitted at the lower end of the nozzle so the mixture falls down. This falling can not be as much controlled as an extrusion, but when the precision in deposition is not required, the falling is not insufficient to do construction.

Presently, there is no deposition mechanism which takes into account. Therefore, this type of process is loosely categorized as extrusion deposition. However, it is not illogical if this type of process is classified as an extrusion deposition process which has the special condition that there is no extrusion taking place. Similarly, a syringe based process where a drop is coming from the syringe without applying much pressure can be classified as a jetting deposition which has the special condition that there is no jetting taking place.

2.6   Field Force

There are some processes that do not require a nozzle or channel to send material for deposition. Material can move from one point to another if there is a force acting on it, this force can be gravitational, magnetic, electric, electrochemical, etc. To take into account the ionic or charge movement, which gives rise to material deposition in AM, a mechanism is required, which is the application of field force.

3   Solid Deposition

Solid deposition (SD) is a family of AM process in which solid materials are used as a feedstock and brought as a solid to a point on or near a platform before they are converted into a desired shape. Feedstocks in the form of powder, wire and rod are used. To transform the feedstock, energy is required, which comes from beam (laser, electron, plasma), arc, heater, friction, or motion.

Examples of SD are:

- Laser beam (l-beam) based: laser engineered net shaping (LENS), laser powder deposition [14], laser wire deposition [15], laser based AM [16]
- Electron beam (e-beam) based: wire e-beam AM (WEAM) [17]
- Arc based: wire arc AM (WAAM)
- Plasma based: plasma welding AM [18]
- Friction based: additive friction stir deposition (AFSD)
- Heater based: fused deposition modeling (FDM), fused pellet modeling (FPM) [19]
- Motion based: cold spray AM (CSAM).

SD is realized through both layerwise and non-layerwise. Non-layerwise type is layerless FDM, while all other SD is layerwise (Fig. 3).

SD can be divided into two broad categories: beam and non-beam based. The former can be divided into three categories: l-beam, e-beam and plasma beam based, while the latter can be divided into four categories: friction, cold spray, arc welding and heater based.

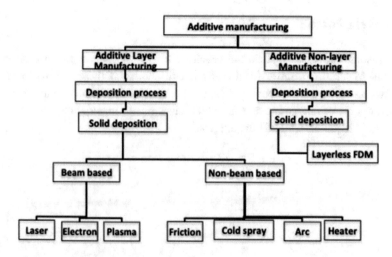

Fig. 3 Classification of solid deposition

4 Beam Solid Deposition

4.1 Advantages

Beam SD has advantages as a beam provides ability to

- Make fine features and parts having high accuracy.
- Scan—otherwise, the scanning needs to be done by moving the machine table or heat source.
- Focus on a spot—a deposition with precision is obtained.
- Defocus—a fast deposition is obtained by covering the increasing area during scanning [20].
- Focus and defocus [21]—beam spot size can be increased and decreased leading to various sizes of deposition.
- Access the remote site of a complex part while repairing [22].

4.2 Disadvantages

Beam has disadvantages: it is not absorbed by many materials, is energy-inefficient, is unsafe, etc. Besides, it is used to melt, which gives disadvantages:

- **Large grain size**—melting layerwise increases the temperature of a build providing conducive environment for a grain to increase in size, causing a decrease in strength.
- **Generation of porosity**—high solidification rate does not let trapped gases emanate, which causes circular pores to form [23].
- **Cracks due to mismatch of thermal expansion**—when a mixture of two materials (e.g. metal and ceramic) is melted, then during solidification, both materials will contract by different amounts depending upon their thermal expansion coefficient, causing stress and possible cracks.
- **Hot cracking**—during solidification, a high melting point material solidifies earlier, leaving a low melting point material still in a liquid state. Cavities or grain boundaries containing such liquid become the source of crack initiation, which brings restriction to the type of materials that can be mixed.
- **Segregation of elements**—depending upon the density and melting point, elements segregate in a melt pool leading to their inhomogeneous distribution and causing a change in mechanical properties.
- **Vaporization of materials**—in order to melt a high melting point material of a mixture of materials, if the temperature of the mixture exceeds the vaporization temperature of

a low melting point material, the low melting point material will vaporize, which will change the mixture composition.

This brings restriction on a maximum melting point difference between any two materials in a mixture. It will narrow the processing window, implying a limited variation in process parameters. If the parameter is scan speed, it will limit the maximum scan speed, slowing the fabrication speed.

- **Anisotropic properties**—a melted layer is cooled through either previously melted layers or a substrate, causing a grain growth in the build direction (columnar grain). This induces two mechanical properties—one in the grain growth direction and another in its transverse direction [24].

5   Laser Solid Deposition

In laser solid deposition (LSD), l-beam coming from a nozzle acts as a heat source, which melts blown powder supplied by the same or different nozzle, or which melts fed wire supplied by the different nozzle. By moving the nozzle parallel to a fixed substrate (or the substrate parallel to a fixed nozzle), a melting point becomes a line. By creating such lines as per the design of a first layer, a physical layer is made.

By moving the nozzle up (or the substrate down), space is created to accommodate 2nd to-be-fabricated layer—this step followed by melting is repeated to fabricate the 2nd layer and so on. The process has variations due to variations in nozzle, feedstock, position of the nozzle, and laser–material interaction.

LSD is an extension of laser cladding, but there is a difference. Laser cladding is meant to apply a coating on a substrate in order to improve its property. The coating conforms to the contour of the substrate. If the coating does not conform, it is not a coating but the application of a coating process to make features.

LSD makes a 3D structure that may be cut off from the substrate to be further used while in laser cladding, the coating is an integral part of the end product and not meant to be cut off from the substrate to be used separately. If LSD akin to laser cladding makes coating on a substrate conforming to its geometry, it is the case of laser cladding process executed in an LSD system.

5.1   Types of Powder Deposition

Coaxial continuous, coaxial discrete, and off-axial powder depositions are practised in LSD.

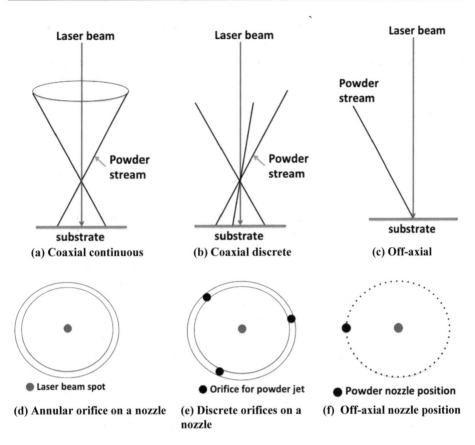

Fig. 4 Schematic diagrams for powder deposition: **a** coaxial continuous, **b** coaxial discrete, **c** off-axial; top views of: **d** annular orifice, **e** discrete orifices, **f** off-axial position

5.1.1 Coaxial Continuous

In this type, powder is deposited around the axis of an l-beam (Fig. 4a). The beam comes from the central area of a nozzle while powders from a continuous annular orifice (Fig. 4d) made at the periphery of the nozzle. The powder stream takes a conical shape around the beam—the axis of the cone is same as that of the l-beam giving the name coaxial deposition. Since the powder is injected from all points at the annular space, the coaxial deposition is coaxial continuous, which is used for fabricating structures with high resolution [25]. As the deposition has no directional dependence, it is suitable to make multi-layer depositions [26].

5.1.2 Coaxial Discrete

In this type, powder is deposited not from all points at the annular space but from selected points (Fig. 4b). Figure 4e shows three discrete points from where powder is injected forming a conical shape (having same axis as that of the beam) around the beam. For less than three discrete points, a conical shape will not form. Discretization helps achieve higher deposition rate but has lower symmetricity of powders' position around the beam in comparison to the former type. Increasing the number of discrete points increases the number of powder jets, and symmetricity. Using nozzles to create powder jets will provide the same effect. This has higher powder divergence than coaxial continuous type, causing a smaller stand-off distance from the substrate. If divergence in coaxial continuous is 8°, it is 15° in this type [27].

5.1.3   Off-Axial

In this type, the axis of a powder stream makes an angle between 0° and 90° with l-beam axis (Fig. 4c, f). At 90°, the stream will not impinge upon a substrate unless either the substrate is tilted or the stream loses its speed and falls. At 0°, the powder jet will no longer be off-axial but coaxial. The powder nozzle is not associated with a beam nozzle, which gives flexibility in setting up an off-axis angle and a stand-off distance for the powder nozzle. This deposition type is used for its simplicity and flexibility [28].

However, this has directional dependence. The relative motion between the nozzle and the substrate may change the deposited powder amount and the deposition site (Fig. 5). The powder stream does not follow a straight line (Fig. 4c) but is bent due to gravity (Fig. 5a, b) [29]. Powders after leaving the nozzle are no longer tied to the nozzle. When the substrate is moving towards them, powders strike the substrate at a nearer location (Fig. 5a) while the substrate is moving away, powders will strike at a farther location (Fig. 5b). This causes the intended strike position to be negatively or positively missed. Changes from the intended position can happen in any deposition process [30].

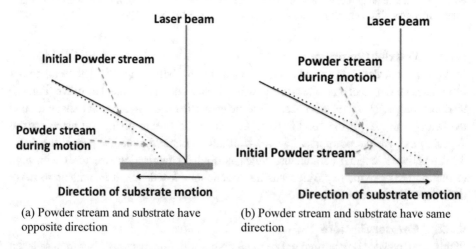

(a) Powder stream and substrate have opposite direction

(b) Powder stream and substrate have same direction

Fig. 5 Directional dependence in off-axial powder deposition, motion in: **a** opposite direction, **b** same direction

This situation can be avoided in two cases: (1) powder jet is injected with a higher speed, which has disadvantage that it may cause ricochet from a solid substrate and create ripples in the melt pool, (2) the substrate moves slow comparing to the feed rate, (3) the powder nozzle is kept closer to the substrate, which may damage the nozzle. If the time for a powder to reach the substrate is small (millisecond), it will not be affected by gravity or drag force [31].

Implications of directional dependence depends on other parameters. For example, if an l-beam spot size is far smaller than a powder stream spot size, and with displacement of the powder stream (due to any motion), the l-beam spot size is always covered by the powder stream spot size, there will be no implication. However, if the l-beam spot is under- or overflowed with powders, these will cause non-uniform deposition.

Employing two powder jets in opposite directions will help—in one direction, one jet will cause deprivation but other will cause overflowing, while in another direction the reverse will happen. Thus, the net effect will be zero [32].

5.2 Laser–Powder Interaction

The timing of interaction between an l-beam and a powder stream influences the process. There are three possibilities as follows:

5.2.1 L-Beam and Powder Stream Interacts Midway

If an l-beam and a powder stream meet on the way before they reach a substrate, it can affect the process. If the powder comes on the way, it will obstruct the l-beam from reaching the substrate. Powder will get heated and the heated powder will reach the substrate, this shadowing effect will bring variation in a melt pool. In a coaxial continuous deposition, the degree of interaction depends upon the geometry of l-beam cone and the cone made by the stream [33] (Fig. 6).

In case the l-beam cone is thinner than surrounded powder stream envelope, the beam will intersect the envelope in a smaller zone (Fig. 6a). Otherwise, the intersection zone will be big (Fig. 6b), which will obstruct the beam more, resulting into more heating of powders. If an off-axial deposition is used, this can be minimized by depositing either at an angle (Fig. 7a) or perpendicular to the substrate in combination with an inclined beam (Fig. 7b).

It can also be minimized in a usual setup (Figs. 4c or 7c)—inclined l-beam [34] though will increase the dissipation of laser energy by increased reflection of the beam. Increasing angle between l-beam path and powder stream path (Fig. 7a) will have minimum shadowing effect, but this decreases both energy-efficiency and powder catchment efficiency (this is the percentage of blown powders ended up in a solidified layer).

If powders, instead of getting just heated get melted, the creation of the melt pool on the substrate is no longer required as a means to melt powders and problems associated

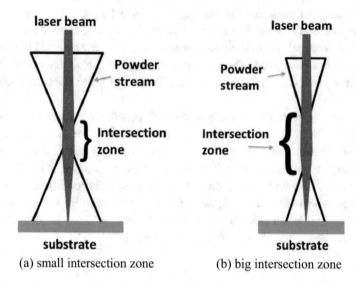

Fig. 6 Intersection between l-beam and powder stream in coaxial continuous deposition: **a** small zone, **b** big zone

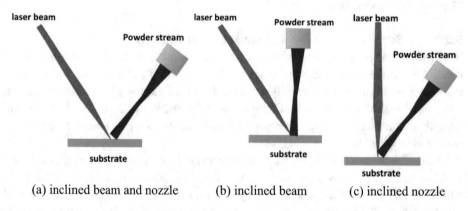

Fig. 7 Minimizing shadowing effect: **a** inclined l-beam and inclined off-axial nozzle, **b** inclined beam and perpendicular nozzle, **c** perpendicular beam and inclined nozzle

with creating and controlling the melt pool will be over, but there are following reasons why powders are not melted:

1. They are not small enough—powders used are from the size of about 20 μm and above, these are not small enough to be melted during small laser–powder interaction time. If these are of sub-micron size, they can be melted. But, using them is not possible as these cannot be fluidized [35] and, therefore, cannot be accelerated using a carrier gas.

During acceleration, they agglomerate and no longer remain of the size of an isolated small powder small enough to be melted.

2. If they are small (depending upon the experimental condition), they vaporise and can form plasma that attenuates the beam. If they are not small, they vaporise partially, which gives rise to recoil pressure to the powder, which changes its direction of flow [36].

3. Laser energy is not sufficient—focal length of the beam is small, which does not let them have sufficient interaction with the beam to gain sufficient energy for melting. This can be overcome by increasing the focal length as well as the laser power. Increasing the focal length, though possible by optics, will bring an additional problem to have a big focused powder stream matching that big length. Increasing the laser power may vaporise the substrate in the course of melting the powder.

Though the problem can be overcome, and liquid drops (by melting powders) will fall on the substrate, which on being continued will give rise to a structure. But the aim of the process is not to facilitate the complete melting on the way but to do the opposite, i.e. to prevent even an isolated case of powder being melted on the way. Besides, there are more economic and energy-efficient ways to create liquid drops than using expensive powder and laser.

5.2.2 L-Beam Follows Powder Stream

Powders are blown on a substrate in order to deposit them, but they will be rebounded from the substrate causing it not to have the intended amount of powders. With an increase in the feed rate and the powder size, rebound force [37] will increase, which will make increasingly difficult to place a powder layer on the substrate to replicate a powder deposition type familiar in powder bed process.

This difficulty is also a source of difference between two processes: bed type and deposition type. Imagining a state where the difficulty disappears—the nozzle is used to additionally create a powder layer that acts as a support to make overhangs on it and there is no need to manoeuvre the axes of a 5-axis CNC machine to circumvent the need of support structure. Though the difficulty will not disappear but can be minimized with the help of an additional nozzle fitted with the sole purpose of placing powders [38].

Though it is impossible to make a powder layer, it is possible to have some powder on a substrate when powders are blown on it. These few powders count when the aim is to create a thin melt pool (of μm thickness), which is desired when minimum dilution (it is a measure of how much substrate material is becoming part of a build) is required. There are a number of strategies (such as decreasing laser power, increasing scan speed, or decreasing feed) to create a thin pool, and these strategies are better than how a thin pool is resulted due to the presence of stray powders. But the point is what will be the consequence if the beam reaches the substrate after the powder has reached.

5.2.3   Powder Stream Follows L-Beam

When an l-beam strikes a substrate, it is supposed to create a melt pool so that when powder stream reaches the substrate, powders will be melted in the pool and the deposition of material will start. How long these powders remain unmelted in the pool determines the surface roughness and build quality. If they remain unmelted for long, there will be no more space for other powders to be accommodated in the pool [39]. If incoming powders can not be accommodated, the build rate will decrease.

Though, powders attenuate the coming beam depending on the configuration of the beam and powder stream, this type of creation of a melt pool and subsequent deposition is practised in multi layer deposition. Size of the pool will increase with an increase in the injection of powders and its temperature will increase with an increase in the injection of hot powders [40].

6   Electron Beam Solid Deposition

Electron beam solid deposition (ESD) is a process in which a part is made by feeding wire in a melt pool created by an e-beam. The movement of a substrate in x- or y-direction with respect to an electron beam makes a layer while layer addition takes place in z-direction (Fig. 8). E-beam loses energy when it ionizes gas molecules present in the environment. To prevent ionization, e-beam must travel in vacuum.

Therefore, the interaction between the wire, substrate and e-beam takes place in a vacuum chamber (Fig. 8). The need of the chamber limits the size of a part that can be made and thus ESD is not having as much freedom as LSD has. Besides, it also limits the type of feedstock that can be used. In the chamber, the deposition of powder is not convenient—it may ionize on the way, small powders may fly because of charging; or carrier gas that brings the powder to the substrate will ionize as well. Therefore, wire is used as a feedstock.

ESD is an energy-efficient process because feedstock is efficient (maximum fraction of the wire fed ends up in deposition) and the energy source is efficient (95% electrical

Fig. 8 Schematic diagram of electron beam solid deposition

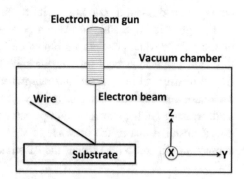

energy is converted into e-beam). There is a limit to the maximum power of the beam that can be used because at high power the X-ray generated due to the beam needs to be shielded from coming out of the chamber.

7   Powder and Wire

The difference between powder and wire as a feedstock is as follows:

7.1   Cost

Powder is more expensive than wire. Many times, powder is produced from wire and therefore the cost of the powder is equal to the cost of the wire plus the cost of conversion of the wire into the powder. Higher cost can also be due to the fact that getting narrow powder size distribution is more difficult than getting uniform wire diameter.

Powder requires carrier gas to be delivered, which increases the total cost. When a big part is fabricated, the cost of feedstock increases the cost significantly, making the wire preferable due to its lower cost [41].

7.2   Availability

Materials that are used to make a wire are limited. The available wire is made mostly from metals. If the metal or alloy is malleable, it is easy to draw wire from such materials, which increases the availability. Powder has higher availability than wire because powder can be made from a higher number of techniques, such as: from solution (joining atom by atom) as well as from big blocks (melting and spraying).

Besides, processing two or three types of powders (metal and ceramics), a wide variety of composite powders (usable in SD) can be formed while for processing two or more materials (metal and ceramic), a limited variety of composite wire (usable in SD) can be formed. This is because the wire needs to be drawn out while powder has no such compulsion, though, majority of the powders should have regular size. If there is irregularity in the diameter of a wire, the whole wire needs to be discarded while if some powders are irregular, these are discarded and the remaining powders can still be used.

7.3   Material Efficiency

During deposition, there are many powders to be controlled while there is just a single wire. Therefore, there is a chance for losing powders while there is no such possibility with the wire. Thus, the powder gives lower deposition efficiency than the wire [42, 43], and a lower building rate [44]. Powders left out during deposition needs to be collected, sorted and recycled, while there is no such problem with the wire. This increases the processing cost with powders as for creating the same size of a part a higher amount of powder is required.

Powder furnishes higher accuracy than the wire because powders can work with smaller melt pools. The wire needs to be thin to give small pool. A thinner wire cannot be fed, which gives limitation to the smaller size of the pool that can be obtained by decreasing the wire diameter. Besides, a small pool cannot be planned to be obtained by partially melting a thick wire as it may cause the wire to be entangled with the build.

Wire is preferred in microgravity because the whole wire can be traced while some powder can spill and pose safety problems [45].

7.4   Processing in Vacuum

In vacuum, wire can be fed with ease without disturbing it. Feeding the powder using carrier gas changes the vacuum pressure due to the presence of gas. Besides, the gas gets ionized, which obscures the e-beam and decreases its efficiency [46].

7.5   Oxidation

For converting a material into a bead, the material needs to be melted. For the same weight, the material in the form of powder rather than wire has a higher surface area. Thus, if the powder instead of the wire is used, a higher surface area is involved, which exposes bigger area to oxidation in an oxidative environment, making powder based than the wire based deposition more vulnerable to oxidation.

7.6   Safety

Powders have higher surface by volume ratio than that of wire. For reactive materials, powders have higher chance to burn. Besides, it is easy to inhale powder, which creates health hazard.

7.7   Effect on Process

Powder gives flexibility to process—a number of types of powders can be mixed and fed through a single nozzle to make a composite [47] while the number of wires that can be fed from a wire-feeder to make composite is limited. For example, ceramic and metal powders can be mixed and fed while feeding ceramic and metal wires together is not equally feasible because of the difference in the stiffness of the ceramic and the metal. In the case of a bundle of ceramic wire and metal wire, both types need to be melted individually to make a composite. If the ceramic wire is not melted, it cannot contribute a ceramic material to the melt pool. Consequently, a ceramic–metal composite will not be formed. In the case of a mixture of ceramic powder and metal powder, even without melting the ceramic, a ceramic–metal composite can be formed.

In wire based deposition, wire needs to be completely melted. If it gets partially melted, it will be attached to the bead causing termination of the process. In powder based deposition, when all powders are expected to be melted and if some are not melted, it causes part imperfections but the process will not be terminated. Hence, feeding excess or insufficient powder can be a strategy for varying properties and making porous structures. Wire feeding does not have such flexibility as wire feed rate needs to correspond with heat input to prevent non-melting of the wire. Therefore, the wire does not offer such variation in properties that can be obtained by a combination of melting and non-melting. It is no wonder that the direction of wire based AM is towards making stronger parts and not a part whose strength is optimized by partially melting the wire. It means wire based deposition does not attempt to make a part consisting of stronger and weaker sections, in which the stronger section is made by full melting of the wire while the weaker section is made by partial melting of the wire.

Wire is generally fed from sideways or off-axial positions while powder can be fed both from sideways and a coaxial position. Wire is confined to a certain point or limited area in a melt pool (Fig. 9) while powder can be defocused or focused to cover all area of the melt pool.

This implies that a wire melts at the certain zone of a melt pool and spreads all over. How far the melted wire spreads depends upon how slow the melt solidifies. If it solidifies earlier, the surface of the melt pool will not be planar, which will decrease surface smoothness. While powder is injected at many places of the melt pool though in smaller quantities in comparison to the bigger volume of the wire tip, melted powder will spread smaller distance (or will not spread) in comparison to the movement of the melted wire.

Fig. 9 Wire feeding at the leading edge of moving melt pool

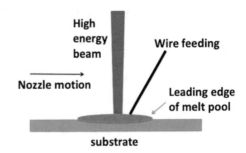

Consequently, melting the powder will not affect surface smoothness as much as melting the wire will. Thus, powder feeding gives better surface finish.

However, all powders need to be melted, there are many powders in comparison to a single wire. With an increase in feed rate, it is more difficult to maintain confinement of all powders than that of a single wire tip in a melt pool. Consequently, from the perspective of melting of feedstock, wire feeding rather than powder feeding allows to attain higher feed rate, leading to a higher deposition rate with wire.

Since wire feeding necessarily entails localization of the wire tip at a particular place in the melt pool, it gives rise to a question—which place of the melt pool is its particular place. A melt pool is created by melting the substrate (or previously melted layer) using a beam, the wire is positioned in such a way that it is to be dipped in the coming melt pool. If the wire is positioned under the beam, it will attenuate the beam. Therefore, any position that is not under the beam is the right place.

For the realization of the process, movement of a beam nozzle (relative to a substrate) or the substrate (relative to the nozzle) is used to create a melt pool. A moving pool thus created has two edges: leading and trailing. Since leading edge is in continuous contact with the beam therefore it is a hot edge while trailing edge is relatively cold and solidifies faster than the leading edge.

Wire fed at the trailing edge has higher chance to solidify (and become part of the solidified melt pool) before the wire acquires enough energy from the melt pool to melt itself and let unmelted feedstock wire detach from the melt pool. The detachment will not hinder the feeding nozzle to move forward parallel to a substrate. This is the reason why the leading edge rather than the trailing edge is the preferred place where the wire is pointed (Fig. 9).

Direction of motion determines which edge of the melt pool is leading or trailing. If the direction reverses, the leading edge will become the trailing edge and vice versa. Therefore, if the position of the wire is not reversed with a reverse in the direction, the performance of the process in both directions will not be same. Changing the position of the wire can be avoided if there are two wire feeders at either side of the beam, and with a change in the direction, alternate wire feeder is used.

7.8   Processing an Inaccessible Area

Wire is a large integrated unit in comparison to disunited powders. It requires certain minimum stiffness to be pointed outside the wire feeder and to reach the melt pool. In the absence of the stiffness, it will not move towards target but will dangle. This stiffness, which is an essential property of the wire, is a problem when the aim is to reach an inaccessible area to do processing.

Powder unlike wire is not an inseparable part of an integrated unit and the moment it leaves nozzle, it no longer remains attached to the nozzle. Therefore, it has more possibility to reach an inaccessible point by other means such as rebounding and falling.

Figure 10 shows an inaccessible area where, at point A, material needs to be deposited. Vertical feeding is not possible due to the geometry of the structure. Because, if the tip of the wire reaches point A, l-beam instead of entering the cavity will be reflected away.

By feeding the wire from an off-axial position (Fig. 11a), the point of repair i.e. A is not accessible. While from the same off-axial position, powder can reach point A by rebounding from the side surface and falling (Fig. 11b). This allows feeding to happen at a melt pool created by the beam and a build to be completed. In the case of the wire feeding, the melt pool thus created will not be able to be fed and the build will not be completed. This type of powder deposition is not an ideal deposition but it shows that what can be achieved with powder cannot be achieved with wire.

This shows powder as a feedstock has an advantage over wire. Though, deposition using wire can still be accomplished by melting the wire in the midway, but the completion of the deposition in this way again conveys the advantage of the powder feeding.

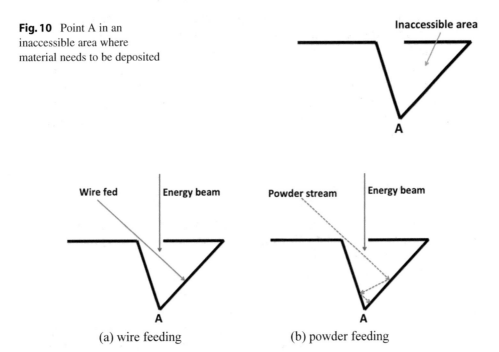

Fig. 10 Point A in an inaccessible area where material needs to be deposited

(a) wire feeding   (b) powder feeding

Fig. 11 Material deposition at point A of an inaccessible area: **a** wire feeding, **b** powder feeding

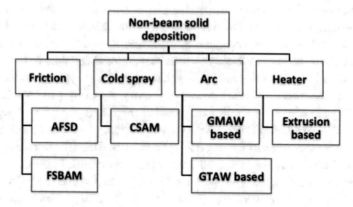

Fig. 12 Classification of non-beam solid deposition

8 Non-beam Solid Deposition

Non-beam solid deposition classified in Fig. 3 is further elaborated in Fig. 12.

Arc based deposition come as an alternative (to beam based deposition) as they do not use expensive beams and can melt higher melting point materials. But they have to deal with melting-induced problems: porosity, residual stress, crack, anisotropy, partial evaporation, etc. Processes based on cold spray or friction do not depend on melting and are, therefore, free from such problems. Besides, there are processes based on filament extrusion, which melt polymers and, therefore, they need a heater and not an arc or beam.

Friction based deposition, which make parts without raising the temperature more than 90% of the melting point of the material, are additive friction stir deposition (AFSD) and friction surfacing based AM (FSBAM) [48]. AFSD is derived from friction stir processing (FSP) while FSBAM is akin to FSP minus stirring.

9 Additive Friction Stir Deposition

In FSP, a rotating pin attached to a shoulder (friction stir tool) stirs a solid material. The pin is an extended tip of a solid cylinder (shoulder). The role of the shoulder is to prevent materials to be extruded out of plane, which is caused due to stirring.

In AFSD, there is no pin while the shoulder has a thorough hole (Fig. 13a). The purpose of the hole is to pass the material onto the substrate from the above, the material can be rod, powder, or small blocks of any shape.

A rod cannot be deposited to a substrate from the above in the same way as a powder through the hole, unless the rod is broken into small pieces. Therefore, a consumable rod, which is rotating and moving, is abraded against the substrate so that it loses material

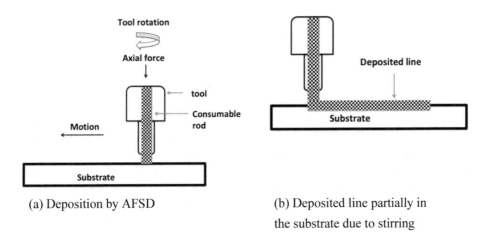

(a) Deposition by AFSD

(b) Deposited line partially in
the substrate due to stirring

Fig. 13 Schematic diagram of additive friction stir deposition

from its surface to the substrate, which is then deposited in the form of a paste (Fig. 13b) [49].

A rotating and moving cylindrical tool deposits powder vertically through its hole. The cross sectional area of the hole is smaller than the non-hole cross-sectional area of the tool. The non-hole section of the tool is called shoulder. When the tool rotates, part of the shoulder encounters already deposited material, which deforms the material before it can escape and become free. Depending on its linear speed, the shoulder encounters the same material times and again, which deforms, mixes and plasticizes it—this is stirring.

A deposited material will be free from this cycle of processing if it is spherical ceramic powder deposited on a ceramic substrate so that rotating shoulder will be able neither to deform and flatten nor to embed the powder onto the substrate. The powder will then roll over through the gap between the shoulder and the substrate, and will eject.

Friction between the shoulder and the substrate raises the temperature that increases the deformation and stirring. The temperature does not exceed the melting point of the material so the benefit gained through solid-state processing is not lost [6]. High rotation speed of the shoulder may induce melting, which will limit the maximum permissible speed, which in turn will limit the fabrication speed.

In the absence of high temperature, there is no driving force for grains to increase in size resulting in small grains responsible for high strength and ductility. Though, if a number of layers are fabricated, long exposure of initial layers to friction-generated temperature causes an increase in the grain size. Stirring and mixing do not let grains settle, resulting in high-angle grain boundaries that are responsible for further increase in the strength.

AFSD applies axial force, an essential component of the process, for material change, or joining. While the force is essential for material transformation, it may deform a complex structure. For making the structure, the tool which applies force, needs to tread over from a thick to a thin support, pass through a gap or over a bridge, or move in a corner. These may require application of less force either for the tool to be able to move or prevent a delicate feature of the structure from collapsing.

Less force implies no material transformation—this brings to a situation where either the material needs to be transformed or the structure needs to built. This situation can be eliminated if no complex structure is built, i.e. simple structure such as rectangular, circular, annular, cylindrical or any such blocks are made. This situation can be partially eliminated if the process is optimized for a range of size of tools so small blocks can be made with small tools and big blocks with big tools, and the design of a structure is chosen, which is the superset of a number of such blocks.

10   Friction Surfacing Based AM

In FSBAM, a rotating consumable rod is heated due to heat generated by inter-frictional force between the rod and a substrate (Fig. 14a). The heating causes the rod material to be plasticized, which is then smeared on the substrate due to an applied axial pressure on the rod (Fig. 14b).

If instead of a rod, a substrate gets heated and plasticized, causing it to flow, the purpose of material deposition by means of the rod will not be served. Therefore, the substrate must satisfy the following conditions:

1. It should be stronger than the rod so the rod plasticizes earlier,

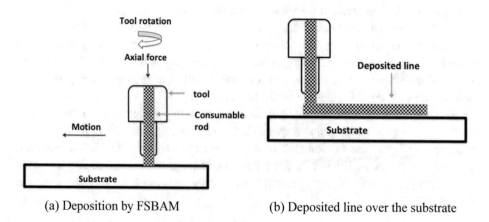

(a) Deposition by FSBAM   (b) Deposited line over the substrate

Fig. 14 Schematic diagram of friction surfacing based AM

2. It should have higher thermal conductivity so the heat is dissipated through it and is not accumulated on it but on the rod,
3. It should be colder. When the rod is hot, it is weaker than the substrate even if both are of the same material (e.g. stainless steel, mild steel, etc.), which ensures that the same material is deposited on the same material (of the substrate), or the same material (of one layer) is deposited on the same material (of another layer) to begin multi-layer fabrication and transition from friction surfacing to AM. This simplifies the role of the substrate. The substrate will only be used for anchoring and supporting fabrication. It is no longer a problem that the material to be deposited is to be compatible with the material of the substrate.

A rod can be heated by frictional heating either by rotating it on some other locations of a substrate or on the surface of another plate [50]. Pre-heating is another option [51]. The hot rod is then brought to the area of the substrate where it needs to deposit (or to be deposited). Cooling the substrate is another method to create a temperature difference between it and the rod. Besides satisfying process requirement, cooling is used to change grain sizes [52].

When a consumable rod begins to deposit, it faces inertia of a substrate. When it exits the substrate, it needs to be detached—this leaves marks on the surface at both entry and exit points, which causes both surface and dimensional inaccuracy. This can be avoided if there is neither an entry nor an exit point, which is possible if processing is planned so both points lie outside the design of a part and can be trimmed away after the fabrication.

The situation, similar to arisen by entry and exit points, will arise again within the design of a part if the rod slows down, takes a turn, or deposits on a feature smaller than its diameter. The bonding between a substrate and a layer or between two layers is a metallurgical bond that is achieved due to forging (axial pressure of rod) on a plasticized material. Thus the properties of the resulting material is similar to a wrought material. The speed of the process depends upon how fast the deposited layer cools down and becomes strong enough for the second layer to be deposited upon.

FSBAM, like AFSD, does not have the means to stir the deposited material, which in turn does not make the material as refined and homogenized as in AFSD. The advantage of the absence of stirring is that small features can be conveniently built without being potentially bent by the stirring action.

11   Cold Spray AM

CSAM is a process in which powder particles are accelerated at high speed to deposit on a substrate to make a structure (Fig. 15). Cold spray means powder particles are not melted, and thus there is no provision for utilizing high temperature sources to cause melting. On the contrary, in thermal spray, partially or fully molten drops are used and thus high temperature sources such as arc, direct current plasma, radio frequency plasma, flame, etc. are required to melt wire, powder or rod to make molten drops.

Cold spray is not a thermal spray, and the thermal spray does not include the cold spray. It does not mean the cold spray has nothing to do with high temperatures. In the cold spray, temperature is increased to 1000 °C, but its purpose is to increase the speed of the gas. It also does not mean the cold spray has nothing to do with the melting—partial melting of powder may happen when the speed of the powder is high and the substrate is at higher temperature—partial melting will provide metallurgical bonding between powders and the substrate or a deposited layer. Thus, the cold spray deposition is not so cold. It, like a thermal spray process, is only not hot enough to be able to melt powders before their deposition.

Cold spray relies on the speed of particles to create bond between the particles and a substrate. Kinetic energy of a particle provides binding energy while in thermal spray, kinetic plus thermal energy provide binding energy. If the speed is high, the particle will deform which will break its oxidized layer exposing its surface to the substrate for a metallurgical bonding to occur. The particle can also get interlocked in the surface roughness or non-uniformity giving rise to a mechanical bond, and can be trapped inside a microcrevice of the substrate surface, resulting in a bond.

If the speed is high, but neither the particle nor the substrate is deformable, the particle will rebound back and there will be no bonding. Thus a combination of high speed and ductility of the material is required. If the substrate is not ductile, by localized heating its non-ductility can be decreased to some extent. If the particle is not ductile, by combining it with another ductile material, ductility of the resulting composite (ductile plus non-ductile material) can be made workable. Thus, a composite powder made from the ductile metal and the non-ductile ceramic can be deposited where the metal will give rise to bonding while the ceramic will impart strength to deposited layers.

A powder which gets deposited without being melted gives certain advantages over a powder that is partially or fully molten during deposition. These advantages are the absence of:

Fig. 15 Schematic diagram of cold spray AM

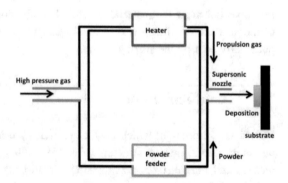

1. Phase transition of materials,
2. Residual stress caused by solidification,
3. Need for heat management,
4. Grain growth, thus nanostructure of the material can be retained,
5. The possibility of deleterious intermetallic compound formation,
6. Splat,
7. Oxidation in presence of an oxidising gas,
8. Need for a processing chamber, therefore no limitation in a part size,
9. Thermal expansion mismatch during multi-material deposition.

Hence, CSAM has advantages over powder fusion based deposition.

11.1 How Cold Spray Is Generated

If high speed gas is available and some powder is injected into it—this is a cold spray. High speed gas coming from a pressurized cylinder or centralized pressurized gas source fulfils one of the basic conditions for cold spray generation. The maximum pressure of the gas thus obtained is limited, which limits the highest speed obtained therefrom. The speed obtained may not be sufficient. If gas is heated on its way, the pressure and the speed obtained can be further increased. The speed obtained again may not be sufficient. This brings a question whether there is a method to increase the speed of the gas without using increasingly powerful cylinders and hotter heaters.

If gas and or gas plus accelerated powders exit through a cylindrical pipe to the substrate, the speed will remain same because the pipe will not increase the speed. If the diameter of the pipe at the end of the cylinder is constricted, the speed will increase if the gas is not compressible. Since there is no gas generated inside a pipe, the amount of gas entering the pipe per unit time must be equal to that leaving per unit time.

The amount of gas that can be accommodated in a pipe depends upon its diameter—if the diameter of the pipe is decreased, it can accommodate less amount of gas. If the flow rate does not increase, the amount of gas entering per unit time will be more than the amount of gas leaving per unit time, and therefore the rate will increase. Therefore, by taking a cylindrical pipe having constricted diameter at one end or taking a conical nozzle the gas speed can be increased without taking resort to extra cylinders or heaters. But, the speed thus achieved will still be limited. Moreover, the gas will diverge, which will increase the diameter of the deposited tracks, which will not be suitable for making small features.

If gas is compressible, the amount of gas that can be accommodated in a pipe will depend not only on the diameter of the pipe but also on the compressibility of the gas. Thus, for a compressible gas, same advantage with a conical nozzle is not obtained.

Therefore, increase in the speed of the compressible gas may not be as high as an increase in the speed of an incompressible gas.

However, if an increase in speed is sufficient enough to increase the speed of a compressible gas more than or equal to the speed of sound, this high speed gas (super sonic gas) increases the compressibility of the gas. Since compressibility increases, the gas no longer follows the principle of an incompressible gas. The principle is that when an incompressible gas passes through a convergent nozzle, its speed increases, and when the same gas passes through a divergent nozzle, its speed decreases.

But, the gas follows the principle of a very compressible gas. The principle is that when a compressible gas passes through a convergent nozzle, its speed decreases, and when the same gas passes through a divergent nozzle, its speed increases. At such supersonic speed resulting in high compressibility, the speed of the compressible gas further increases when it enters a divergent nozzle or when it exits from a convergent section to a divergent section of a convergent-divergent nozzle.

Thus, a gas coming from a compressed cylinder gets its speed increased in the convergent section of a convergent-divergent nozzle or de laval nozzle because the gas is slightly compressible (behaving like an incompressible gas), and the speed gets further increased in the divergent section of the de laval nozzle because the gas is no longer just slightly compressible but has become highly compressible because of a change in the behaviour of the gas at such high speed.

Thus, using the right type of nozzle is another method to increase the speed of a gas. When a particle is injected in this gas, the particle is accelerated at high speed and can be used for cold spray AM.

12   Arc Based Deposition

Classification of arc based deposition is given in Fig. 16. Arc created between two electrodes is used as a heat source in metal welding. This technique of metal welding is extended to create 3D structures. The resulting AM processes are arc based deposition. The processes are based on gas tungsten arc welding (GTAW) (Fig. 17), plasma welding (plasma non-transferred arc, plasma transferred arc) (Fig. 18), and gas metal arc welding (GMAW) (Fig. 19). Wire arc AM (WAAM) is any type of arc based deposition which uses wire feedstock.

12.1   What Is Arc

When voltage is applied between two electrodes, air molecules or atoms present between electrode tips get polarized, which causes positive and negative charges of an atom gets separated, creating positive and negative poles within the atom. If the voltage is increased

Fig. 16 Classification of arc based deposition

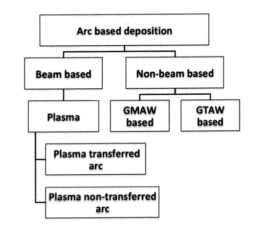

Fig. 17 Schematic diagram of gas tungsten arc welding based deposition

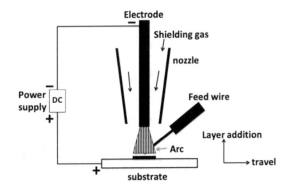

further, these poles are not confined within the atomic boundary but gets detached from each other. After detachment, they are not poles but charges—this conversion of non-charged entities into charges is ionization.

The air between tips consists of charges that are collectively called plasma. The charges move towards the electrodes, which gives rise to an arc that is a continuous line of charges between electrodes. The movement faces obstacles due to the presence of other particles in the air. While overcoming the obstacles, friction occurs between moving charges and particles, which gives rise to heat. Thus, the creation of an arc can entail the creation of a heat source. Higher voltage means more ionization, stronger attraction and more intense arc giving more heat. Higher current means more injection of electrons in the plasma, resulting in more ionization and a stronger arc.

Ionization not only depends on voltage and current but on the type of a gas present between the electrodes. Gas having lower ionization potential will be ionized at a lower voltage than that having higher ionization potential. Thus, argon gas (15.7 eV) unlike helium gas (24.5 eV) is easier to be ionized, which makes argon gas suitable to be used at a lower voltage.

Fig. 18 Schematic diagrams of plasma welding based deposition: **a** non-transferred arc, **b** transferred arc

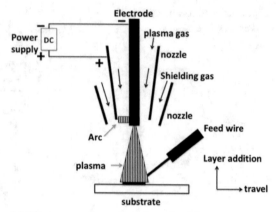

(a) Plasma non-transferred arc based

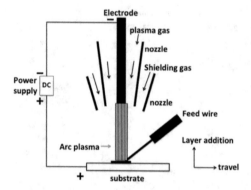

(b) Plasma transferred arc based

Fig. 19 Schematic diagram of gas metal arc welding based deposition

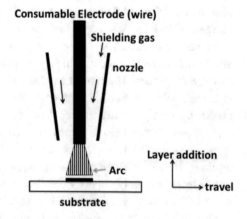

The type of the gas not only decides an amount of ions in an arc but also whereabout of heat in the arc. If the gas has high thermal conductivity, e.g., helium gas (0.151 W/mK), the heat will not be confined within the arc line but will spread out making the influence of the arc wider, resulting in a wider welding bead. If the gas has low thermal conductivity such as argon gas (0.018 W/mK), the heat will not spread, which will not result in a wider welding but, possibly, a deeper welding. For creating a deeper welding, arc should contain high heat that is not possible by just having one favourable property, i.e. low thermal conductivity of the gas.

If the gas in the arc furnishes extra heat, it will serve the purpose. Oxygen gas during ionization splits into atoms and furnishes extra heat (dissociation energy). Thus, the presence of oxygen will contribute to the arc intensity. If a small amount of oxygen is mixed with argon, the intensity will increase. Thermal conductivity of oxygen (0.027 W/mK) is slightly higher than that of argon, but it will not offset a gain in the heat intensity (by taking heat away) obtained due to the dissociation of oxygen.

Heat generated in an arc affects an electrode, which can cause it to be melted and used as a source of material for AM. Though the heat can degrade it when it is not supposed to be melted but is supposed to act only as an electrical terminal to create the arc. In GTAW based AM, tungsten electrode acts as an electrical terminal, cooling of which is one of the ways to mitigate heat-induced degradation. Cooling happens when a gas flows by the electrode. This is the same gas that is a precursor for the arc. If the gas has high thermal conductivity, the heat transfer will be more and the electrode will cool more. Therefore, helium, which has higher thermal conductivity cools electrodes more than argon does.

When direct current is flowing and tungsten electrode is maintained as a positive electrode, the degradation is fast because positive tungsten ions from it are pulled by a negative electrode. It gives rise to the loss of atoms from the tungsten electrode resulting in its depletion. Using alternating current instead helps avoid this type of degradation. When it is maintained as a negative electrode, it does not have any such degradation through losing positive ions. Besides, heat is generated more at the positive than at the negative electrode. Therefore, for the same setting of current and voltage, the positive electrode becomes hotter and degrades more.

Bigger positive ions and generally smaller negative ions (electrons) are constituents of an arc. When voltage is applied across electrodes, negative ions being smaller move fast towards the positive electrode while positive ions being bigger move slow towards the negative electrode. Fast electrons have higher impact on the positive electrode making it hotter.

Thus, a tungsten electrode that is maintained as a negative electrode is cooler and safer. Its efficiency is further improved by alloying it with thoria. When voltage is applied across electrodes, it is the negative electrode from which electrons enter into the arc, the transfer of electrons from electrode to the arc depends also upon the crystal structure of the electrode. When tungsten is mixed with another material, which decreases its work function, the transfer of electrons becomes easier and the efficiency of the electrode increases. Addition of 2 wt.% thoria decreases its work function from 4.5 to 2.6 eV.

12.2 Gas Tungsten Arc Welding Based

The process, as the name implies, uses a tungsten electrode to create an arc, while the prefix 'gas' implies a shielding gas used to prevent oxidation of materials melted during welding. Shielding gas flows around the arc, which shields atmospheric gas from coming near to the arc and prevents the reaction between the atmospheric gas and arc–metal interaction zones (Fig. 17).

Shielding gas is generally an inert gas (Ar or He) but can also be a mixture of inert and non-inert gases. Since it is the same gas from which an arc is formed, it serves the dual purpose of shielding and arc-creating. Besides, the arc type depends upon the gas type. Therefore, its selection does not depend upon the sole purpose of shielding but on the arc type needed.

To create an arc, tungsten electrode is made cathode and a work-piece or substrate is made anode when direct current is used. Since 70% of heat is generated at the anode, this type of polarity helps over-heat the workpiece and under-heat the cathode. Since the aim is to melt workpiece (or the material on it) rather than the tungsten electrode, this polarity helps conserve energy by channelizing the heat near the workpiece.

The material in the form of powder or wire is fed in order to be melted either directly by an arc or indirectly by a melt pool that is created by the arc. Since the stability of the arc remains largely undisturbed by its effect on melting, it gives an advantage for controlling the deposition.

12.2.1 Plasma Welding Based

Arc created in GTAW exists near an electrode and thus operational ability of the arc is limited by its physical proximity to the electrode. If the arc were created far away from the electrode, it could have more freedom—the effect of melting due to the arc could have been planned without worrying about its negative consequence on the electrode. Besides, feeding could not have been restricted due to a constricted space between the electrode and the workpiece, and the size of the melt pool could not have been limited due to a limited variation in arc dimension. These limitations are in sharp contrast to the freedom provided by an l-beam—the effect of an l-beam spot on a workpiece does not affect a laser source.

Limitations imposed by an arc can be overcome by creating better heat sources utilizing the arc. As the arc is hot, if a gas passes through it, the gas becomes hot, which can be used as a heat source. For obtaining it, a tungsten rod is fitted inside a hollow nozzle, and while the rod usually acts as a cathode, the nozzle acts as an anode. When the gas is flown through the nozzle, the gas provides a medium to create an arc between the rod and the nozzle. Besides, the gas coming out from the nozzle is affected by the arc that is already created, i.e. the gas becomes hot and ionized, which is plasma. The plasma can reach away from the location of the cathode and therefore, its effect can be realized at a farther distance. The reach of the plasma depends upon the gas speed, the nozzle type and the arc temperature.

This new heat source is due to thus generated plasma and the arc struck between a nozzle and a cathode. To produce this, the location of the arc does not need to be changed. It means the arc remains maintained between the nozzle and the cathode before the flow of plasma gas (through the nozzle to the workpiece) and after its flow. If the arc remains maintained, the arc is not transferred anywhere. Therefore, this new heat source is called plasma non-transferred arc (Fig. 18a).

What if an arc is no longer maintained between the nozzle and the cathode, and the arc is transferred to the workpiece? Then, the arc striking and heating the workpiece is no longer the same as the plasma striking and heating the workpiece. This method of heating the workpiece can give rise to one more type of a heat source. In this method, the arc is first created between the nozzle and the cathode. Since this arc does not serve or intend to serve either directly or indirectly as a heat source, a less hot arc or low-current arc is sufficient, which is started between the nozzle and the cathode.

When the gas passes through it, and the plasma strikes between the cathode and the workpiece, it is easier to create another arc between the cathode and the workpiece. Thus, when polarity of the workpiece is changed from neutral to positive with respect to the tungsten rod, another arc is started. The initial arc created serves its purpose when this new arc is created, the initial arc is then extinguished. The initial arc is named as pilot arc while the heat source due to this new arc is called plasma transferred arc (Fig. 18b). Pilot arc is also started by a high frequency generator and in that case, the polarity of the nozzle and the tungsten electrode periodically changes.

Plasma transferred arc needs to pass through a constricted nozzle orifice and is therefore thin resulting in the concentration of heat energy. Since plasma gas also passes through the orifice, the gas is converted into plasma due to its interaction with the arc, which results in a heat source consisting of both arc and plasma. Therefore, this heat source has higher temperature (almost two times) than the arc in GTAW does have. Besides, in plasma transferred or non-transferred arc, shielding gas is separately used from plasma gas, which provides an advantage that the striking of the arc is not affected by the shielding gas as much as it is affected in GTAW.

12.3 Gas Metal Arc Welding Based

During the creation of an arc between two electrodes, when one electrode is melted and ready to be deposited, the need for feeding materials is eliminated. In GMAW, out of two electrodes, one electrode is consumable and other electrode is a workpiece or substrate (Fig. 19). If one electrode is consumable, it means it is a feedstock.

Since the form of a consumable electrode is usually rod or wire, the feedstock is usually rod or wire. It gives a more efficient transfer of heat from an arc to the wire than a process where the electrode and the wire are not the same—this results in a fast deposition.

A consumable electrode provides some disadvantage. If the electrode gets consumed, arc length will change. This change will disturb the process by changing either voltage or current, which will require further adjustment of the voltage, current or feed rate.

The process makes a structure by transferring molten parts of an electrode to a work-piece (Fig. 19). The transfer depends mainly upon how much molten the molten part is. If it is little molten, mainly at a condition of low-current, this will not be detached from the electrode on its own and therefore, no deposition will occur. If it is little more molten, mainly at a medium-current condition, this will be detached on its own, resulting in a deposition.

To be detached, gravitational force needs to overcome surface tension. After detachment, droplets which are around the size of the electrode diameter will be deposited. If it is well molten, mainly at a high-current condition, it will not only be detached on its own but will take the shape of small droplets. Because at this condition, there is high heat energy that decreases the surface tension and therefore, a small gravitation force due to the droplet will be sufficient to overcome the surface tension.

With an increase in current, heat energy increases, which increases the temperature of molten material allowing it to overcome surface tension and get fragmented into droplets. Instead of increasing the current to impart heat, it can also be imparted by other methods such as the use of oxidizing gas. These three conditions of current show three types of respective material transfer: short-circuit transfer related to low-current, globular transfer related to medium-current and spray transfer related to high-current.

In a short-circuit transfer, since material is not detached on its own, it needs to be detached by initiating a short-circuit between an electrode (wire) and a work-piece. When material is not detached from the wire tip, the wire is continuously fed, which continuously decreases the arc length between the wire and the workpiece until the length or gap becomes zero—this is short-circuit. A constant current then passes from the wire to the workpiece, which gives rise to resistive heating at the interface between the wire and the workpiece, this heat melts the tip.

Molten material detaches from the tip by overcoming the surface tension at the tip, this again creates an arc between the tip and the deposited material, which brings an end to the short-circuit.

Thus, in this type of material transfer, a continuous cycle of creation (of the short-circuit) and end of the creation gives rise to a continuous deposition that enables fabrication. In this type, an uncontrolled resistive heating produces spatter which decreases the precision of the deposition.

To avoid resistive heating and reduce spatter, a variation of this type of metal transfer, cold metal transfer, is developed. The prefix 'cold' denotes the absence of (resistive) heating. Thus, current is set to zero and the wire is retracted the moment the semi-molten tip of the wire contacts the workpiece, and thus resistive heating is avoided. In the absence of the resistive heating, the tip does not get heat energy from the resistive heating and needs to melt without it. Thus, a low-current is able to melt the tip but is not high enough to detach the molten material from the tip on its own. Thus, cold metal transfer is able to furnish 3D steel structures with high deposition rate and reproducibility [53].

13   Extrusion Based Deposition

Extrusion in extrusion based deposition implies extrusion of material in a continuous shaped form. The shaping of the material after it is out from a nozzle and before it reaches a substrate gives the process its name. This shaping distinguishes the process from other processes (such as ink jetting) [54] those also rely on nozzles but to send liquid drops through without extruding.

The shaping is not solely due to a nozzle but due to materials, temperature, environmental conditions, applied force, etc. But the role of the nozzle has more ability to separate extrusion based deposition from other AM processes.

Extrusion based deposition using solid polymers are FDM, fused filament fabrication [55], FPM [56], fused layer modeling [57], big area AM [58], powder melt extrusion [59], composite extrusion modeling [60]. Ceramic based extrusion deposition are fused deposition of ceramics (FDC) [61], robocasting [62], direct writing [63], etc. Besides, IJP using extrusion [1], and liquid metal using extrusion [2] are other extrusion based deposition.

13.1   Solid Polymer Feedstock Type

Filaments, pellets and powders are used as feedstock. Filament based deposition is FDM (Fig. 20a) while FPM uses pellet. Powder based deposition is powder melt extrusion and composite extrusion modeling (Fig. 20b).

Filaments are widely used feedstock and almost all commercial systems use them while few systems use pellets. The advantage of a filament is that it works as not only a feedstock but also a machine part of a FDM system because a moving filament acts as a piston to push molten filaments though nozzle.

Thus, a filament brings simplicity in AM system development. This is one of the reasons why filaments based systems are in abundance. But if the filament is to act as a piston, it must possess more properties than required by a feedstock that needs to act

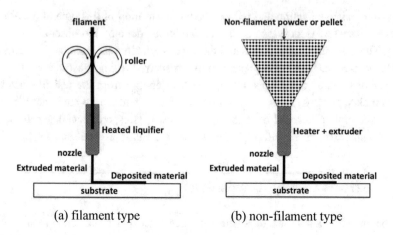

(a) filament type   (b) non-filament type

Fig. 20 Schematic diagrams of extrusion based deposition using: **a** filament, **b** non-filament

only as a source of materials. This is the disadvantage, which demands extra effort in its development and places extra conditions on a material to be suitable to become a filament.

A filament should have enough rigidity that it will not buckle when it pushes the molten material, but it should not have much rigidity that it will not bend when it needs to be driven in a non-straight path from its source. It should not be too weak to break when it pushes the material through. Moreover, it should not be made so strong that making it stronger means increasing its melting point, which in turn will cause difficulty in melting. Thus, it needs to satisfy conflicting property requirements, which excludes many materials to be converted into filaments and then into parts.

If there is an extrusion based deposition that uses feedstock other than filaments, the process will not be deprived of the advantage gained from extrusion as well as be free from the disadvantage of using filaments. If feedstocks such as pellets and powders are used, there will be a difference in how they are brought to a print head. They will need a different setup for storing the feedstock, and a screw or piston for extruding. But there will be no difference after they are melted and extruded. Using them provides opportunity to use various polymers and their mixtures that otherwise would not have been possible as various combinations of materials do not allow a filament to be drawn. This has facilitated to use elastomeric, ceramic and metallic materials in extrusion based deposition. A metallic product can also be formed by using a composite feedstock made from metal and polymer and, in a downstream process, removing the polymer from the extruded part [64].

13.2   Layerless Fused Deposition Modeling

In FDM, during the formation of a layer, the height of a nozzle from the layer does not change. The nozzle moves up after the last line of the layer is deposited, i.e. it gets opportunity to move up only after the completion of each layer.

What if a nozzle does not wait for the completion of a layer before it moves up. In layerless FDM, the nozzle constantly moves up while it deposits a line (Fig. 21). Thus, simultaneously moving up and forward by the nozzle makes a deposition path spiral [65]. Moving up in such a way does not let a layer be made. If there are no layers, there does not exist any problem associated with joining them.

Constantly upward moving nozzle means there will always be an empty space, i.e. the process does not have capability to fill up an empty space that is made by a spiral path. For example, for making a thin cylinder by a spiral path, there will always be space at the centre, i.e. the formation of the cylinder will necessarily create a cylindrical space at the centre of the cylinder along its height.

A solid thin cylinder can be made if extruded material has low viscosity, so the material is not able to retain itself within a deposited line and will move inwards to fill up a gap at the centre. But that material will also move outwards increasing inaccuracy. A solid inaccurate thin cylinder will thus be made, which is not due to the merit but demerit of the process.

For making a big solid cylinder, the nozzle has again to go down and start moving up, the movement of the nozzle will be obstructed by the structure it has already

Fig. 21 Schematic diagram of layerless fused deposition modeling

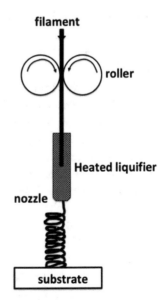

made. The process is thus not able to make a solid part but only hollow parts. ALM has disadvantages, but unlike this process, it provides a route to make a solid part.

If a process makes only hollow objects such as empty cylinders, the process is not different from FDM that already makes such cylinders. Without support structures, layerless FDM can make empty cylinders with increasing or decreasing diameter, resulting in conical or inverted conical type structures, respectively. Thus it has more scope than FDM to change diameters and vertical angles of conical structures.

In FDM, a deposited line is partially supported by the previously deposited line. Depending upon a change in diameter, the next deposited line will slip, which is prevented by creating support structures. In layerless FDM, there is a continuous deposited spiral line that provides lateral support and prevents the structure from falling. The process thus demonstrates a method to avoid support structures [66].

14   Layerless Deposition and Layerless Bed Process

Layerless FDM is a process that is an application of FDM to create spiral structures. Layerless FDM and FDM are different because both have different process steps. The process step that is different is how extrudates are deposited. In layerless FDM, it is deposited non-layerwise while in FDM, it is layerwise—when FDM is applied to create a structure for which same process steps are not followed, a new process is formed.

Thus, a layerwise process, FDM, gives rise to another process, layerless FDM. For creating a new process, what FDM does is not something that cannot be done by other processes. Any deposition process can deposit material in a spiral way, and a new process will be ready. It has already been done by other processes, e.g. WAAM [67], solvent-cast direct writing (DW) [68] (in DW, solvent of extrudate ink is evaporated, causing the extrudate to retain its shape).

There is no need for a nozzle moving in a circular path to go up if the platform is going down. There is also no need for the nozzle either to move or to go up if the platform is simultaneously rotating and moving down. Creating a relative spiral movement between the platform and the nozzle gives a new process belonging to a different process group, and a spiral structure.

This happens in bed process when platform rotates and goes down [69], and in photopolymer bed process (in inverse orientation) when the platform rotates and goes up [70].

In powder bed process, there is no nozzle to deposit materials and therefore the bed process has to arrange materials without the help of a nozzle. Thus, a material hopper is required to spread the material and create the bed. Thus both processes (bed and deposition) requires material to be deposited. One process does it with a nozzle while another process does it with a hopper. When a spiral structure is made, both processes look to be closer to each other, blurring the division between them. Though powder is deposited in both cases, there is a difference. One process requires just the deposition with a nozzle. Another process requires deposition plus spreading. Though spreading is not much but there is no absence of spreading. That's why instead of a nozzle, a powder hopper is required, which is having an attached coater with it. Therefore, creating a layerless structure does not help the bed process become a deposition process.

Why a spiral structure is made? It is made because for fabrication of certain geometries, support structures are not needed when a part is made spiral way. Support structures are needed more in a deposition process than in a bed process because of lack of a bed in the deposition process [71]. Therefore, fabrication through spiral structure has more utility in a deposition process when a part necessarily requires a support structure in either process, and a bed in the bed process provides the necessary support.

What if the bed in the bed process does not provide the necessary support? It means when a part is made in either process, it requires the creation of special support structures in both processes. Even then, a spiral deposition process is preferred to a spiral bed process when a spiral based process needs to be developed only to overcome the need for support structures. It is because the working of the spiral deposition process does not require any modification in the existing deposition system provided the system has enough degrees of freedom for its nozzle to move in both horizontal and vertical planes. While existing powder bed systems, especially commercial systems, are not fit for spiral based process and require extensive modification in terms of platform movement and material feeding.

However, for photopolymer, a spiral deposition process does not have much edge over a spiral bed process as there exist systems for inverse stereolithography that can be modified for the spiral bed process.

15 Comparing Friction Based with Fusion Based Deposition

- **Small features**: In friction based deposition, fabrication of a small feature is possible if it withstands tool pressure during fabrication, while in fusion based deposition, there is no tool pressure and the smallest features depend upon the minimum dimension of the beam or arc.
- **Material properties**: Fusion based deposition gives material properties similar to a cast material, which can differ due to the thermal gradient, and the direction and rate of solidification, while friction based deposition gives properties similar to a wrought material [49].
- **Grain types**: Fusion based deposition is versatile, which provides grains of various types, which can be columnar, single crystal, equiaxed of various sizes, and a mixture of all grain types within a given area. By changing the scanning parameters and substrate temperatures, different types of grains are obtained at various heights from a substrate.

Friction based deposition provides equiaxed grains. Therefore, if the property of a part is not adequate because other processes could not furnish equiaxed grains, the part can be fabricated in this process using feedstock of any grain types as the process converts any grain types into an equiaxed type.

- **Processing an inaccessible area**: A beam can reach far away than a friction tool can therefore, fusion based deposition is better when modification needs to be done or a small feature needs to be added in an inaccessible area.
- **Tool wear**: Only in friction based deposition does a tool wear out needing replacement.
- **Flexibility**: Fusion based deposition can make features of various sizes while friction based deposition do not have such flexibility due to fixed tool sizes.
- **Retaining the original crystal structure**: If the aim is to retain the original crystal structure of feedstock, friction based deposition because of being solid-state and low-temperature process is a better alternative.
- **Feedstock external property**: Fusion based deposition, for operational reason, emphasizes external feedstock property such as shape, size, surface roughness, and dimensional accuracy. For examples, if wire is not of uniform diameter, it cannot be fed with a consistent rate; if powder has wide size distribution, it will not give uniform amount of deposition.

Friction based deposition is free from such stringent requirements. For example, powders of any size can be fed through hollow cylindrical tool. This has led to even experimenting with any size of chips as a feedstock [72]. Absence of stringent requirement means the low cost of feedstock, which makes the process inexpensive [6].

16   Air Deposition

The classification of air and ion deposition is given in Fig. 22.

Air deposition is mainly Aerosol jetting (AJ), which is also known as aerosol jet printing [73], and is used to fabricate small features, mainly for electronic applications [74].

Aerosol implies that a liquid or solid particle is suspended in air. In AJ, liquid particles are used, which become constituents of a part made while the air acts as a carrier to transport particles and facilitate the process. If particles are suspended in the air, they are not big but are the size of a nano or micrometer. If such small particles are transported, there will be a small amount of particles per second ready to be deposited to make a part, which is suitable to make only small parts from ~10 μm up to 1 mm (~100 nm resolution).

Most of the parts are used along with substrates on which they are made. The process is thus used more for modifying a substrate and adding value to it than making a part to

Fig. 22 Air and ion deposition

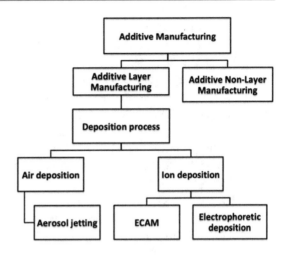

be removed from it. Its applications are making structures such as antenna, interconnects, electronic circuits, conductive lines either on planar or non-planar surfaces, etc.

For making a structure, AJ requires aerosol and its subsequent deposition (Fig. 23). The creation of the aerosol requires the creation of liquid droplets and mixing them with air. The creation of droplets requires agitation of liquid so the separation of the droplets from the bulk liquid can take place. The agitation is possible by vibrating the liquid placed in a container so that some liquid particles on the surface of the liquid overcome surface tension and get separated from the bulk to form droplets. This is also the concept of ultrasonic atomization where ultrasonic vibration is transferred to the liquid through another liquid medium.

The agitation is also possible by impacting the liquid with a gas moving at a high velocity, generally coming from a compressed source. Such gas moving parallel to a liquid surface and brushing the surface will detach some liquid from the surface and turn it into droplets. For a high-viscous liquid, higher velocity is required. This requirement can be overcome by heating the liquid to make it internally agitated. Because a gas moving at the same velocity gives increased amount of droplets when impacted on internally

Fig. 23 Schematic diagram of aerosol jetting

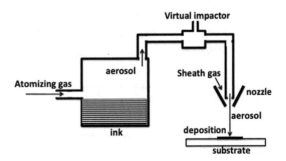

agitated liquid. Heating as a method to facilitate droplet formation is generally used when the liquid kept still in a container needs to be atomized. If a high velocity gas moves perpendicular to a liquid stream coming from a nozzle, the collision of the gas with the liquid stream creates droplets. This method is used for atomizing liquid streams.

Liquid droplets thus created will fall and get lost if these are not carried away. The gas, which creates them, called an atomizing gas, can also carry them. The gas can then be called a carrier gas. After atomizing, droplets of various sizes are created. Bigger size (or of higher inertia) droplets will require higher velocity carrier gas in order to be transported. If the atomizing gas is a carrier gas, the option for increasing or decreasing the velocity is limited. After the creation and transportation of aerosols, and before their deposition, some mechanism is required to exclude droplets of extreme (high and low) sizes. Such droplets are not suitable if a high resolution is required.

When aerosol moves in a pipe, the presence of an exhaust or outlet in the pipe may let some gas (containing aerosol) move out to atmosphere. The amount of the gas moving out depends upon the exhaust. If the size of the exhaust is big or the exhaust flow rate is high, more gas will move out resulting in a loss of smaller size droplets. Besides causing the loss of droplets, the exhaust causes a decrease in the gas velocity. If the gas velocity decreases, it will not be able to carry big size droplets that will be dropped from the gas. Thus, using an exhaust eliminates both smallest and biggest droplets from the gas. This mechanism or controlling device is called virtual impactor that eliminates extreme size droplets because of their lowest or highest inertia.

Carrier gas deprived of droplets of extreme sizes while containing of medium size is directed towards a nozzle to be deposited through it. If the gas is sent through the nozzle, the size of the gas flow diameter will increase with an increase in the distance of the nozzle from the substrate. Hence, the gas will not provide patterns having high definition and density.

To collimate the carrier gas, another gas called sheath gas is used. The sheath gas flows in the direction of the carrier gas flow, and covers it with an aim to converge at some point. If the shape of the carrier gas is just like a solid cylinder, the sheath gas is a hollow converging cone surrounding the cylinder. By collimating, the sheath gas helps constrict the deposition.

If there are several carrier gases (containing different materials) flowing at the same velocity and originating from different atomizers, it is possible to merge these flows into one and make a multi-material deposition. Changing the amount of materials in one gas flow by changing the setting of atomizer will provide variations in the deposited material.

17   Ion Deposition

There are two ion deposition processes: (1) electrolytic solution based, and (2) colloidal solution based.

When an external potential difference is applied across solution using electrodes, charged particles or ions move through the solution—positive ions (cations) move towards negative electrode (cathode) while negative ions (anions) move towards positive electrode (anode).

Movement of ions towards electrodes leads to accumulation of materials on electrodes, which is akin to deposition of materials on a substrate in AM—this is a method used to add materials to an ionic solution based AM. This method is not new, it is already practised in electroplating or electrophoretic deposition for making thin coatings that are 2D structures. The method is also applied in electroforming to make 3D structures [75].

In electroforming, materials are deposited on electrodes of various shapes, these electrodes are called tool or mandrel. On the removal of the electrode from the deposited material, the remaining hollow structure is a desired 3D structure. This structure is dependent upon tools [76]. The dependency limits the process to fabricate only those structures that conform to the shape of tools. The present AM process is free from such limitations.

Ionic solution implies a liquid either containing ions (e.g. NaCl solution) or having potential to be ionized when an external voltage difference is applied. When potential is applied across the ionic solution such as copper sulphate solution, copper ions move giving rise to copper plating. This happens in electroplating and the solution is called electrolytic solution. AM based on these types of solutions is electrolytic solution based AM, also known as electrochemical AM [77] (Fig. 24).

A colloidal solution may also be an ionic solution as it contains charged particles suspended in liquid. These charges remain present in the solution even in the absence of an external applied electric field. These charges move in the presence of an external field and give rise to a deposition—this happens in electrophoretic deposition, and the ionic solution is a colloidal solution. AM based on these types of solution is colloidal solution based AM, also known as electrophoretic deposition based AM [78].

17.1 Electrolytic Solution Based

When metal ions move and get deposited on a cathode, they form a layer on it. If the cathode is a big plate and anode is a small rod facing small area of the cathode, the formation of the layer will be limited to that area (Fig. 24a). Exact size of the layer will depend upon variables: gap between two electrodes, throwing power of the electrode, applied voltage, form of the voltage, insulation on the side surface of the anode, concentration of electrolytes, etc. But changing the size of the anode provides a way to localize the formation of layers or deposition of materials at some selected area [79, 80].

If the anode moves parallel to the cathode, the deposited layer moves and creates a layer in the form of a line having width corresponding to the width of the anode (Fig. 24b). If the anode moves away from the cathode, the deposited material starts growing away from the cathode. If there is more growing, a pillar is made on the cathode (Fig. 24c).

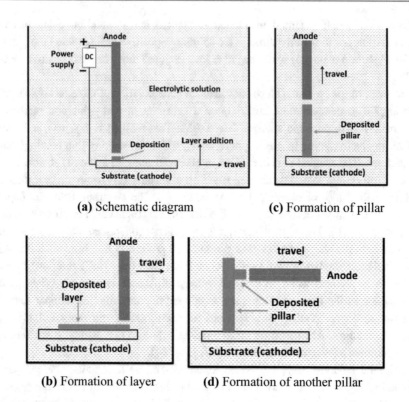

(a) Schematic diagram   (c) Formation of pillar

(b) Formation of layer   (d) Formation of another pillar

Fig. 24   Electrolytic solution based deposition: **a** diagram, **b** layer, **c** pillar, and **d** another pillar

After the formation of the pillar, if the anode again moves parallel to the cathode, another pillar will form on the earlier pillar at right angle to it (Fig. 24d). With a change in the direction of the movement of the anode, many such pillars can be formed. This will result in a structure similar to a pillar having many arms—these are overhang structures that, in other AM processes, require either support structures or a change in the orientation of the geometry [81], but in this process, these are formed without such requirement [82].

The addition of materials happens ion by ion, and when these ions are getting added on the side surface of a pillar to make an arm, these have no such possibility to succumb to gravitational force and collapse as happens in a drop by drop AM process. It can collapse if the addition does not furnish enough strength to the structure, but the structure will not collapse for the reason that the ions could not be placed on the side surface. While in IJP or FDM, it is not possible to initiate the formation of a structure as liquid drops or extrudates can not be placed on the side surface.

The process gives opportunity to add materials wherever it is required leading to a method to make a complex part [83]. Hence, materials can be added anywhere on a pillar to make arms at various angles. If layerwise fabrication is followed, the pillar can be

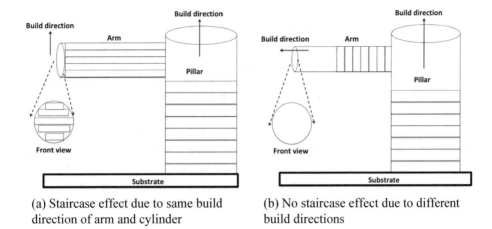

(a) Staircase effect due to same build direction of arm and cylinder

(b) No staircase effect due to different build directions

Fig. 25 Effect of build directions in layerwise process

fabricated well, but the arm will show gaps between two layers (Fig. 25a). Thus, during the fabrication of the arm, the disadvantage of the layerwise fabrication will be visible.

If, after the fabrication of the pillar, the tool does not follow layerwise fabrication in that build direction, there will not be any disadvantage due to the layerwise fabrication. After the layerwise fabrication of the pillar—the pillar acts as if it is a new cathodic plate, the fabrication of the arm becomes the fabrication of the pillar and the tool still follows the layerwise method—the resulting structure (arm) will not show any gap because the layerwise fabrication changes the build direction by 90° (Fig. 25b).

The present example of a pillar and an arm shows how the disadvantage of layerwise method can be overcome by changing the build direction. But it will require software that will slice a CAD model both in horizontal and vertical directions [84].

Commercial AM systems have one build direction [85]. Though, changing the build direction during a build is doable—(1) after fabrication of a pillar, if the position of the cathodic plate is changed by 90° and the build is continued, (2) after fabrication of a pillar, a new CAD model is used for fabricating the arm, the model is sliced vertically instead of horizontally. A new tool path is created, which gives a build direction that is at 90° to the old build direction that was used to create the pillar, the CAD model is positioned at the position of the arm of the old CAD model and the build is continued. Both options show it is possible to manufacture one pillar and arm.

But what if the part is not so simple, the part does not consist of just one pillar and arm but has hundreds of pillars and arms at different angles, the above options will still work but will require process planning.

The problem associated with changing a build direction increases when layerwise fabrication needs to be maintained while changing the direction. One of the reasons for increase is that a layer is inflexible. If the size of a part is big or the size of its cross-section is big, the size or perimeter of the layer will be big. Perimeter of a layer is inflexible and constant, it comes along with a 3D model. If there is any flexibility, it is in the thickness of a layer, which can be varied. In the case of a change in the orientation of a part having unequal dimensions, the perimeter can be changed but is still decided by the model of the part, and is therefore inflexible during fabrication.

In layerwise fabrication, it is the layer that is the basic building block—it is not because layer is not made from smaller blocks but because unless a complete layer is fabricated, the fabrication does not progress to the next layer. There is no such possibility as to delay the fabrication of the fraction of a layer till next two or three layers are formed.

One of the problems of layerwise fabrication is that the layer perimeter is always big—it does not imply that the method is only meant to make big parts—it implies that the layer size is always very big in comparison to the size of the tool that creates the layer. The size of the tool means anode size in this process, laser spot size in LPBF, e-beam spot size in EPBF, liquid drop size in IJP, or extrudate diameter in FDM.

What if the layer size is equal to the tool size, there will be no need for scanning, there will be no more problem for finding the right overlap between adjacent scanned lines and no need to be careful to integrate every bit of such a big layer, there will also be no need to search right parameters to join such big layers. Thus, fabricating with a layer size equal to the tool size gives advantages over fabricating with a big layer size.

In a usual layer-tool setting, the layer by virtue of being a layer is big while the tool by virtue of being a tool is small—these sizes are two extremes. By selecting a basic building block smaller than and other than a layer will provide a compromise between two extremes.

A layer has some finite thickness and is similar to a rectangular plate that can be considered as an assembly of many smaller cubes or volumes that are similar to a cube-like shape. These cubes make a layer and then make a 3D model. These cubes may not make a layer and then make a 3D model. A cube may look like a layer, and a layer may look like a cube. But, the difference between a cube and a layer as a basic building block is build direction. An assembly of layers has only one fixed build direction that is perpendicular to all layers since an assembly of layers can have only one normal passing through all layers.

An assembly of cubes has only one fixed build direction that is perpendicular to all cubes, if a cube is a layer. An assembly of cubes can have many build directions. Since a cube has six faces, it can have a maximum of six directions moving away from six faces. Since fabrication happens on a substrate or a platform, there can not be any build direction towards the substrate resulting in a maximum of five build directions. It is not usual but also not impossible to have a case of six build directions where fabrication will take place without using a substrate—such as in two photon polymerization where fabrication happens in the middle of liquid and can be proceeded in any direction.

A 3D model, consisting of such cubes or cube type elements also referred as voxel, can have many build directions, if algorithm is well developed, and will not be affected by the limitations of layerwise fabrication. Using voxels instead of layers in an example of pillar and arm, there will be no more need for orienting the cathodic plate or using more than a single CAD file to overcome the limitations of layerwise fabrication.

17.2 Colloidal Solution Based

Deposition of metal ions or using salt solutions does not allow many types of materials to be deposited. A colloidal solution provides versatility in materials choice as materials of any types such as polymers, alloys, ceramics are available in the form of colloids. These unlike metal ions can be of either charges depending on the types of additives and solutions, which can be deposited on either electrodes termed as cathodic or anodic electrophoretic deposition [86]. Voltage is applied across electrodes to drive these colloidal charged particles while in the case of ionic solution based AM, voltage is applied to ionize the solution and drive ions.

A solvent is chosen such that it will not ionize readily. If it ionizes, ions will change pH of the solution and disturbs the stability of colloids, or ions will instead be deposited and interfere with the deposition of charged colloidal particles. The particles need to be smaller than one μm so their positions in the solution will not be affected by gravity. If particles are bigger, they will either sediment or be sedimenting.

If they are sedimenting, the solution will have fewer particles on the upper side of the solution while more particles on the bottom side. It will result in a thinner deposition on the top while a thicker deposition on the bottom of a vertical electrode. Thus, bigger particles will not provide a deposition of uniform but graded thickness. Since, addition happens atom by atom, the process is slow making it suitable for small parts.

Conductivity of electrode is important for uniform deposition to occur. If the conductivity is low, there will be poor deposition on the electrode. This fact can be utilized to accomplish the deposition at a particular area on a low conductive electrode by increasing the conductivity of that particular area. It is increased by irradiating a photoconductive electrode with light.

If a photoconductive electrode plate is used, it can be irradiated on its back side to create deposition on its front side that is in contact with colloidal solution. If the irradiation makes a pattern, the deposition of charged particles will make a pattern on the front side. Thus, a 3D structure though of limited geometry can be created [78].

18   Liquid Deposition

IJP [87], digital ink jet printing [88], direct inkjet printing [89], polymer jetting, photopolymer jetting (Fig. 26), rapid freeze prototyping (RFP) etc. are liquid deposition (LD).

If the material is not liquid, it is melted in order to be deposited. If it has high melting point, e.g. ceramic, it is mixed in a carrier liquid. If it is a high molecular weight polymer, it is dispersed in a carrier liquid to make it low-viscous so that it can be ejected through a nozzle.

The liquid should not be solidified either in the nozzle or after leaving it, and should reach a substrate in the state of liquid. Besides, the liquid composition should not change due to chemical reaction or physical segregation or evaporation. It can reach in the form of either continuous stream or drops. But, the continuous stream has propensity to break before it reaches, therefore, the liquid deposition is a drop wise deposition. The process can thus be controlled by controlling the ejection of drops.

Drops are detached from the bulk of liquid through nozzle by the application of: electric field [90], variable magnetic field [91, 92], vibration, sound waves, heating, vapour pressure, or force using physical objects such as piston or screw, etc. The number of drops ejected is synchronized with the speed of the nozzle relative to the substrate so that the speed should not be either high causing a gap or low causing an overlap between two deposited drops.

When a drop reaches a substrate, the condition of the drop depends upon the impact [93]. If it has high impact, the drop will be flat, and its shape will be elliptical. The high impact will cause splash leading to a loss of material from the drop. The elliptical drop may retract if it has high surface tension [87]. If it has low surface tension, the drop will expand.

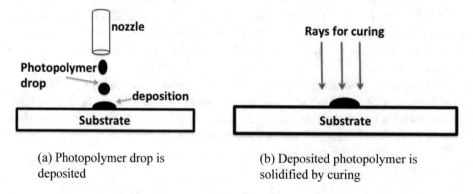

(a) Photopolymer drop is deposited

(b) Deposited photopolymer is solidified by curing

Fig. 26 Schematic diagram of photopolymer jetting: **a** material jetted, **b** material solidified

When successive drops reach the substrate, the drops coalesce and then lose their shapes to make a line. Solidification of drops or lines depends upon the phase transformation, polymerization (Fig. 26b) and gelling, or vaporization of carrier liquid. It is facilitated by changing environment on the substrate by localized heating or curing using beam.

If there are many nozzles, many such adjacent lines can be made to make a layer, and thus the fabrication can be expedited. If there are many nozzles using many materials, there will be many lines made from different materials, which can provide a technique to make a multi-material part. If different materials intend to serve different functions, the part can work as a multi-functional part.

Making a multi-material part requires compatibility between different materials so their ejection from the nozzles will not cause mismatch in ejection times and drop sizes. Besides, there should not be a large difference between inter-material (drops from different nozzles) coalescence and intra-material (drops coming from the same nozzle) coalescence. If drops coming from different nozzles are of different colours, a multi-colour part can be made [94].

18.1 Water Deposition

For making a part from water, e.g. an ice structure, water deposition needs to be controlled and the substrate needs to be kept at a sub-zero temperature so the moment water touches the substrate, it solidifies (Fig. 27).

When an ice structure is made on a substrate and the structure is growing, it will start to melt because the substrate temperature though maintained at liquid nitrogen ($-140\ °C$) will not be able to cool it. This requires the whole setup of a nozzle and a substrate to be kept inside a cold chamber ($-20\ °C$) so the structure will not melt [95]. But, keeping inside may affect the water either within the nozzle or in transit to the substrate, which will cause water to start to solidify. Therefore, the nozzle is kept near the substrate. The water is ejected using more than a critical pressure so the flow of water is not obstructed by icing on the nozzle. Besides, the flow will break off the icing [96].

Fig. 27 Schematic diagram of rapid freeze prototyping

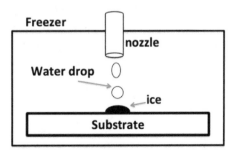

Two nozzles are required for making a complex structure having an overhang: one for the main material and the other for the supporting material. One nozzle can do the job of two nozzles: first, it will deposit the main material and the next time, it will deposit the supporting material as per need—but then the same nozzle needs to be fitted with different liquid sources periodically, which will delay the fabrication, contaminate the liquid, change the heat transfer time and compromise the accuracy.

When two materials are used: one is water and the second is the supporting material of lower freezing point than water (e.g. NaCl or brine solution), then after fabrication, supporting material structure is removed by placing the part in a chamber having temperature more than the freezing point of brine and less than the freezing point of water. This will melt the supporting structure and not the main structure.

In the absence of two materials, what will happen if water does the job of two materials: water can do the job of a supporting material—the supporting structure is made weaker so it is removed by the application of force—this method is used in other AM processes without damaging the main structure [71], but this may not be as convenient since the removal is done inside a cold chamber.

The ice structure thus fabricated will melt away if removed from the cold chamber and therefore, its use needs to be worked out before it is removed. If the ice structure after fabrication is used to make a non-ice structure, then the non-ice structure can be removed from the cold chamber for further application. The example of such non-ice structure is a ceramic shell made from an ice structure, i.e. ice pattern while the application is investment casting.

The process is applied to make an ice pattern for investment casting. In comparison to a polymer pattern [97], the ice pattern gives an advantage as it does not need to be burnt to remove it from a ceramic shell but is removed by bringing the pattern (covered with the shell) outside the chamber.

To create a ceramic shell, the ice pattern is dipped in a ceramic slurry. The slurry should not be frozen so it requires to be free from water, and thus a special slurry is required. The slurry coated on the pattern dries with the help of a catalyst and the shell thus formed from the slurry is further strengthened by heat treatment [98].

Ice parts made from this process have not many applications if possible applications in the area of making ice sculptures are not counted, but the method learnt is applied in an area of tissue engineering where a scaffold not printed in a sub-zero environment will be affected by porogens. Thus, scaffolds are printed using this method by replacing water with a biological solution—this process is named cryogenic prototyping [99].

19   Slurry Deposition

In slurry deposition, slurry is deposited from a nozzle, which helps make ceramic and metal components (Fig. 28). If slurry is not used, there are no other ways but by melting to make big metal components. Melting high melting point materials requires extra effort. For example, it requires to control microstructure development and crack mitigation, devices for melting, and dedicated systems for heat management—these all increase cost as well [100].

There are expensive systems that are powder or wire based requiring l-beam or e-beam. Melting low melting point materials such as a tin alloy [101] or an aluminum alloy [102] is not so difficult. There are processes available for depositing these materials [103], but parts made are weak and can not be a substitute for high strength parts. Hence, theses processes are incapable as they can not be used for processing high melting point materials. Slurry deposition is a solution to these problems—it is neither as difficult as a deposition process related to high melting point materials nor as incapable as a process related to low melting point materials.

Slurry deposition can be a substitute for a deposition process related to high melting point materials but slurry is not a substitute for these materials. Slurry contains these materials that can be maximum around 30 vol.% in IJP [89, 104, 105], 60 vol.% in 3D gel printing [106]. Lower the amount means lower disturbance to the original process, e.g. if solid content in ink jetting is just 2 vol.%, the process parameters may not be much different than that required for pure ink, thus less obstacle to the ink flow. But such low amount does not help get strong parts and is not suitable for many applications.

Slurry extruded on a platform should not be deformed on its own weight and should sustain the form when another layer is deposited on it. Higher solid loading provides strength to an extruded material but requires high force to extrude. The high force may break the continuity of the extruded material, which will cause defects in part fabrication [107].

Due to the presence of binder or gel present in a slurry, extruded materials are bonded with other extruded materials or a previously deposited layer before they solidify. Gel is preferred in some materials (tricalcium silicate) because it prolongs solidification of

Fig. 28 Schematic diagram of slurry deposition

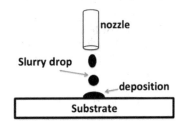

extruded filaments, which gives filaments sufficient time to bond. Besides, gel increases flow property of the slurry, which helps it to be extruded uniformly [108].

References

1. Zheng S, Zlatin M, Selvaganapathy PR, Brook MA (2018) Multiple modulus silicone elastomers using 3D extrusion printing of low viscosity inks. Addit Manuf 24:86–92
2. Gannarapu A, Arda Gozen B (2019) Micro-extrusion-based additive manufacturing with liquid metals and alloys: flow and deposition driven by oxide skin mechanics. Extreme Mech Lett 33:100554
3. Izadi M, Farzaneh A, Mohammed M et al (2020) A review of laser engineered net shaping (LENS) build and process parameters of metallic parts. Rapid Prototyp J 26(6):1059–1078
4. Chaturvedi M, Scutelnicu E, Rusu CC et al (2021) Wire arc additive manufacturing: review on recent findings and challenges in industrial applications and materials characterization. Metals 11(6):939
5. Rajan K, Samykano M, Kadirgama K et al (2022) Fused deposition modeling: process, materials, parameters, properties, and applications. Int J Adv Manuf Technol 1–40
6. Yu HZ (2022) Emerging processes—friction stir based. In: Caballero FG (ed) Encyclopedia of materials: metals and alloys. Elsevier, pp 153–161
7. Castrejon-Pita JR, Baxter WRS, Morgan J et al (2013) Future, opportunities and challenges of inkjet technologies. At Sprays 23(6)
8. Pathak S, Saha GC (2017) Development of sustainable cold spray coatings and 3D additive manufacturing components for repair/manufacturing applications: a critical review. Coatings 7(8):122
9. Brant A, Sundaram M (2022) Electrochemical additive manufacturing of graded NiCoFeCu structures for electromagnetic applications. Manuf Lett 31:52–55
10. Kumar S (2020) Other solid deposition process. In: Additive manufacturing processes. Springer, Cham, pp 111–130
11. Grong O, Sandnes L, Berto F (2019) A status report on the hybrid metal extrusion & bonding (HYB) process and its applications. Mater Des Process Commun 1(2)
12. Blackburn S, Szymiczek M (2021) Extrusion. In: Pomeroy M (ed) Encyclopedia of materials: technical ceramics and glasses. Elsevier, pp 162–178
13. Blindheim J, Grong O, Welo T, Steinert M (2020) On the mechanical integrity of AA6082 3D structures deposited by hybrid metal extrusion & bonding additive manufacturing. J Mater Process Technol 282:116684
14. Vilar R (2014) Laser powder deposition. In: Comprehensive materials processing, vol 10. Elsevier Ltd, pp 163–216
15. Huang W, Xiao J, Chen S, Jiang X (2020) Control of wire melting behavior during laser hot wire deposition of aluminum alloy. Opt Laser Technol 150:107978
16. Yan Z, Liu W, Tang Z et al (2018) Review on thermal analysis in laser-based additive manufacturing. Opt Laser Technol 106:427–441
17. Tarasov SY, Filippov AV, Shamarin NN et al (2019) Microstructural evolution and chemical corrosion of electron beam wire-feed additively manufactured AISI 304 stainless steel. J Alloys Compd 803:364–370
18. Feng Y, Zhan B, He J, Wang K (2018) The double-wire feed and plasma arc additive manufacturing process for deposition in Cr-Ni stainless steel. J Mater Process Technol 259:206–215

19. Nieto DM, López VC, Molina SI (2018) Large-format polymeric pellet-based additive manufacturing for the naval industry. Addit Manuf 23:79–85
20. Steen WM, Majumder J (2010) Laser material processing. Springer-Verlag London Limited
21. Wang L, Zhu G, Shi T et al (2018) Laser direct metal deposition process of thin-walled parts using variable spot by inside-beam powder feeding. Rapid Prototyp J 24(1):18–27
22. Chen H, Lu Y, Luo D et al (2020) Epitaxial laser deposition of single crystal Ni-based superalloys: repair of complex geometry. J Mater Process Technol 285:116782
23. Zhao T, Wang Y, Xu T et al (2021) Some factors affecting porosity in directed energy deposition of AlMgScZr-alloys. Opt Laser Technol 143:107337
24. Shamsaei N, Yadollahi A, Bian L, Thompson SM (2015) An overview of direct laser deposition for additive manufacturing; part II: mechanical behavior, process parameter optimization and control. Addit Manuf 8:12–35
25. Oliveira UD, Ocelík V, De Hosson JTM (2005) Analysis of coaxial laser cladding processing conditions. Surf Coat Technol 197(2–3):127–136
26. Eisenbarth D, Esteves PMB, Wirth F, Wegener K (2019) Spatial powder flow measurement and efficiency prediction for laser direct metal deposition. Surf Coat Technol 362:397–408
27. Gao X, Yao XX, Niu FY, Zhang Z (2022) The influence of nozzle geometry on powder flow behaviors in directed energy deposition additive manufacturing. Adv Powder Technol 33(3):103487
28. Vilar R (1999) Laser cladding. J Laser Appl 11(2):64–79
29. Turichin G, Zemlyakov E, Klimova O, Babkin K (2016) Hydrodynamic instability in high-speed direct laser deposition for additive manufacturing. Phys Procedia 83:674–683
30. Yuan L, Pan Z, Ding D et al (2021) Fabrication of metallic parts with overhanging structures using the robotic wire arc additive manufacturing. J Manuf Process 63:24–34
31. Wu J, Zhao P, Wei H et al (2018) Development of powder distribution model of discontinuous coaxial powder stream in laser direct metal deposition. Powder Technol 340:449–458
32. Kumar S (2020) Beam based solid deposition process. In: Additive manufacturing processes. Springer, Cham, pp 93–109
33. Li L, Huang Y (2018) Interaction of l-beam, powder stream and molten pool in laser deposition processing with coaxial nozzle. J Phys Conf Ser 1063:012078
34. Meacock C, Vilar R (2008) Laser powder microdeposition of CP2 titanium. Mater Des 29:353–361
35. Geldart D (1973) Types of gas fluidization. Powder Technol 7:285–292
36. Sergachev DV, Kovalev OB, Grachev GN et al (2020) Diagnostics of powder particle parameters under laser radiation in direct material deposition. Opt Laser Technol 121:105842
37. McLaskey GC, Glaser SD (2010) Hertzian impact: experimental study of the force pulse and resulting stress waves. J Accoust Soc Am 128(3):1087–1096
38. Kumar S (2020) Future additive manufacturing processes. In: Additive manufacturing processes. Springer, Cham, pp 187–202
39. Haley JC, Schoenung JM, Lavernia EJ (2019) Modelling particle impact on the melt pool and wettability effects in laser directed energy deposition additive manufacturing. Mater Sci Eng A 761:138052
40. Pirch N, Linnenbrink S, Gasser A, Schleifenbaum H (2019) Laser-aided directed energy deposition of metal powder along edges. Int J Heat Mass Transf 143:118464
41. Hassen AA, Noakes M, Nandwana P et al (2020) Scaling up metal additive manufacturing process to fabricate molds for composite manufacturing. Addit Manuf 32:101093
42. Schmidt M, Merklein M, Bourell D et al (2017) Laser based additive manufacturing in industry and academia. CIRP Ann 66(2):561–583

43. Kumar P, Jain NK (2020) Effect of material form on deposition characteristics in micro-plasma transferred arc additive manufacturing process. CIRP J Manuf Sci Technol 30:195–205

44. Blinn B, Lion P, Jordan O et al (2021) Process-influenced fatigue behavior of AISI 316L manufactured by powder-and wire-based laser direct energy deposition. Mater Sci Eng A 818:141383

45. Watson JK, Taminger KMB, Hafley RA, Petersen DD (2002) Development of a prototype low voltage electron beam freeform fabrication system. In: SFF proceedings, pp 458–465

46. Taminger KMB, Hafley RA (2013) Electron beam freeform fabrication: a rapid metal deposition process. In: Proceedings of the 3rd annual automotive composites conference, Troy, MI

47. Kumar S (2022) Comparison. In: Additive manufacturing solutions. Springer, Cham, pp 57–92

48. Dilip JJS, Babu S, Rajan SV et al (2013) Use of friction surfacing for additive manufacturing. Mater Manuf Process 28:1–6

49. Schultz JP, Creehan KD (2014) Friction stir fabrication. US patent US 8893954 B2

50. Rao KP, Sankar A, Rafi HK (2012) Friction surfacing on nonferrous substrate: a feasibility study. Int J Adv Manuf Technol 65(5–8):755–762

51. Gandra J, Krohn H, Miranda RM et al (2014) Friction surfacing—a review. J Mater Process Technol 214(5):1062–1093

52. Mishra RS, Ma ZY (2005) Friction stir welding and processing. Mater Sci Eng R 50(1–2):1–78

53. Ali Y, Henckell P, Hildebrand J et al (2019) Wire arc additive manufacturing of hot work tool steel with CMT process. J Mater Process Technol 269:109–116

54. Chi B, Jiao Z, Yang W (2017) Design and experimental study on the freeform fabrication with polymer melt deposition. Rapid Prototyp J 23(3):633–641

55. Brenken B, Barocio E, Favaloro A et al (2018) Fused filament fabrication of fiber-reinforced polymers: a review. Addit Manuf 21:1–16

56. Wang Z, Liu R, Sparks T, Liou F (2016) Large scale deposition system by an industrial robot (I): design of fused pellet modeling system and extrusion process analysis. 3D Print Addit Manuf 3(1):39–47

57. Kumar N, Jain PK, Tandon P, Pandey PM (2018) Investigation on the effects of process parameters in CNC assisted pellet based fused layer modeling process. J Manuf Process 35:428–436

58. Roschli A, Gaul KT, Boulger AM et al (2019) Designing for big area additive manufacturing. Addit Manuf 25:275–285

59. Boyle BM, Xiong PT, Mensch TE et al (2019) 3D printing using powder melt extrusion. Addit Manuf 29:100811

60. Lieberwirth C, Harder A, Seitz H (2017) Extrusion based additive manufacturing. J Mech Eng Autom 7:79–83

61. Bellini A (2002) Fused deposition of ceramics: a comprehensive experimental, analytical and computational study of material behavior, fabrication process and equipment design. Drexel University

62. Feilden E, Blanca EGT, Giuliani F et al (2016) Robocasting of structural ceramic parts with hydrogel inks. J Eur Ceram Soc 36(10):2525–2533

63. Mondal D, Willett TL (2020) Mechanical properties of nanocomposite biomaterials improved by extrusion during direct ink writing. J Mech Behav Biomed Mater 104:103653

64. Singh G, Missiaen JM, Bouvard D, Chaix JM (2021) Additive manufacturing of 17–4 PH steel using metal injection molding feedstock: analysis of 3D extrusion printing, debinding and sintering. Addit Manuf 47:102287

65. O'Dowd P, Hoskins S, Geisow A, Walters P (2015) Modulated extrusion for textured 3D printing. In: NIP & digital fabrication conference, vol 1, pp 173–178

66. Kanada Y (2015) Support-less horizontal filament stacking by layer-less FDM. In: SFF proceedings, pp 56–70

67. Gu X, Hou Z, Xu J et al (2017) A novel additive manufacturing method for spiral parts. In: 2017 IEEE 7th annual international conference on CYBER technology in automation, control, and intelligent systems, pp 791–796

68. Guo S, Gosselin F, Guerin N et al (2015) Solvent-cast three-dimensional printing of multifunctional microsystems. Small 9(24):4118–4122

69. Hauser C, Sutcliffe C, Egan M, Fox P (2005) Spiral growth manufacturing (SGM)—a continuous additive manufacturing technology for processing metal powder by selective laser melting. In: SFF symposium proceedings, Texas, USA

70. Dudley K (2015) 3D printing using spiral buildup. US patent US20140265034A1

71. Kumar S (2022) Fabrication strategy. In: Additive manufacturing solutions. Springer, Cham, pp 111–144

72. Jordon JB, Allison PG, Phillips BJ et al (2020) Direct recycling of machine chips through a novel solid-state additive manufacturing process. Mater Des 193:108850

73. Goh GL, Agarwala S, Tan YJ, Yeong WY (2018) A low cost and flexible carbon nanotube pH sensor fabricated using aerosol jet technology for live cell applications. Sens Actuators B Chem 260:227–235

74. Wilkinson NJ, Smith MAA, Kay RW et al (2019) A review of aerosol jet printing—a non-traditional hybrid process for micro-manufacturing. Int J Adv Manuf Technol 1–21

75. Castellano PMH, Vega ANB, Padilla ND et al (2017) Design and manufacture of structured surfaces by electroforming. Procedia Manuf 13:402–409

76. Matsuzaki R, Kanatani T, Todoroki A (2019) Multi-material additive manufacturing of polymers and metals using fused filament fabrication and electroforming. Addit Manuf 29:100812

77. Kamraj A, Lewis S, Sundaram M (2016) Numerical study of localized electrochemical deposition for micro electrochemical additive manufacturing. Procedia CIRP 42:788–792

78. Mora J, Dudoff JK, Moran BD et al (2018) Projection based light-directed electrophoretic deposition for additive manufacturing. Addit Manuf 22:330–333

79. Habib MA, Gan SW, Rahman M (2009) Fabrication of complex shape electrodes by localized electrochemical deposition. J Mater Process Technol 209(9):4453–4458

80. Lin JC, Chang TK, Yang JH et al (2010) Localized electrochemical deposition of micrometer copper columns by pulse plating. Electrochim Acta 55(6):888–1894

81. Paul R, Anand S (2015) Optimization of layered manufacturing process for reducing form errors with minimal support structures. J Manuf Syst 36:231–243

82. Brant A, Sundaram M (2016) A novel electrochemical micro additive manufacturing method of overhanging metal parts without reliance on support structures. Procedia Manuf 5:928–943

83. Manukyan N, Kamaraj A, Sundaram M (2019) Localized electrochemical deposition using ultra-high frequency pulsed power. Procedia Manuf 34:197–204

84. Singh P, Dutta D (2001) Multi-direction slicing for layered manufacturing. J Comput Inf Sci Eng 1(2):129–142

85. Coupek D, Friedrich J, Battran D, Riedel O (2018) Reduction of support structures and building time by optimized path planning algorithms in multi-axis additive manufacturing. Procedia CIRP 67:221–226

86. Pikalova EY, Kalinina EG (2019) Electrophoretic deposition in the solid oxide fuel cell technology: fundamentals and recent advances. Renew Sustain Energy Rev 116:109440

87. Derby B (2015) Additive manufacturing of ceramic components by ink jet printing. Engineering 1(1):113–123

88. Lee JH, Kweon JW, Cho WS et al (2018) Formulation and characterization of black ceramic ink for a digital ink-jet printing. Ceram Int 44:14151–14157

89. Cappi B, Özkol E, Ebert J, Telle R (2008) Direct inkjet printing of Si_3N_4: characterization of ink, green bodies and microstructure. J Eur Ceram Soc 28(13):2625–2628
90. Ball AK, Das R, Das D et al (2018) Design, development and experimental investigation of E-jet based additive manufacturing process. Mater Today Proc 5:7355–7362
91. Jayabal DKK, Zope K, Cormier D (2018) Fabrication of support-less engineered lattice structures via jetting of molten aluminum droplets. In: SFF symposium proceedings, pp 757–764
92. Simonelli M, Aboulkhair N, Rasa M et al (2019) Towards digital metal additive manufacturing via high-temperature drop-on-demand jetting. Addit Manuf 30:100930
93. Wijshoff H (2018) Drop dynamics in the inkjet printing process. Curr Opin Colloid Interface Sci 36:20–27
94. Meisel N, Dillard D, Williams C (2018) Impact of material concentration and distribution on composite parts manufactured via multi-material jetting. Rapid Prototyp J 24(5):872–879
95. Leu M, Isanaka SP, Richards VL (2009) Increase of heat transfer to reduce build time in rapid freeze prototyping. In: SFF symposium proceedings, pp 219–230
96. Barnett E, Angeles J, Pasini D, Sijpkes P (2009) Robot-assisted rapid prototyping for ice structures. In: 2009 IEEE international conference on robotics and automation, Kobe
97. Kumar S (2022) Application. In: Additive manufacturing solutions. Springer, Cham, pp 93–110
98. Zhang W, Leu MC (2000) Investment casting with ice patterns made by rapid freeze prototyping. In: SFF symposium proceedings, pp 66–72
99. Pham CB, Leong KF, Lim TC, Chian KS (2008) Rapid freeze prototyping technique in bioplotters for tissue scaffold fabrication. Rapid Prototyp J 14(4):246–253
100. Sames WJ, List FA, Pannala S et al (2016) The metallurgy and processing science of additive manufacturing. Int Mater Rev 1–46
101. Vega EJ, Cabeza MG, Muñoz-Sánchez BN et al (2014) A novel technique to produce metallic microdrops for additive manufacturing. Int J Adv Manuf Technol 70:1395–1402
102. Zuo H, Li H, Qi L, Zhong S (2016) Influence of interfacial bonding between metal droplets on tensile properties of 7075 aluminum billets by additive manufacturing technique. J Mater Sci Technol 32(5):485–488
103. Fang X, Wei Z, Du J et al (2017) Forming metal components through a fused-coating based additive manufacturing. Rapid Prototyp J 23(5):893–903
104. Özkol E, Ebert J, Uibel K et al (2009) Development of high solid content aqueous 3Y-TZP suspensions for direct inkjet printing using a thermal inkjet printer. J Eur Ceram Soc 29(3):403–409
105. Hagen D, Kovar D, Beaman JJ, Gammage M (2019) Laser flash sintering of additive manufacturing of ceramics. ARL-TR-8657, Defence Tech Info Centre, US
106. Ren X, Shao H, Lin T, Zheng H (2016) 3D gel-printing—an additive manufacturing method for producing complex shaped parts. Mater Des 101:80–87
107. Tang S, Yang L, Li G et al (2019) 3D printing of highly-loaded slurries via layered extrusion forming: parameters optimization and control. Addit Manuf 28:546–553
108. Wu W, Liu W, Jiang J et al (2019) Preparation and performance evaluation of silica gel/tricalcium silicate composite slurry for 3D printing. J Non-Cryst Solids 503–504:334–339

Printed in the United States
by Baker & Taylor Publisher Services